Material and Process Design for Lightweight Structures

Material and Process Design for Lightweight Structures

Special Issue Editor

Talal Al-Samman

MDPI • Basel • Beijing • Wuhan • Barcelona • Belgrade

MDPI

Special Issue Editor
Talal Al-Samman
Institute of Physical Metallurgy and Metal Physics (IMM)
Germany

Editorial Office
MDPI
St. Alban-Anlage 66
4052 Basel, Switzerland

This is a reprint of articles from the Special Issue published online in the open access journal *Metals* (ISSN 2075-4701) from 2018 to 2019 (available at: https://www.mdpi.com/journal/metals/special_issues/material_process_design_lightweight_structures).

For citation purposes, cite each article independently as indicated on the article page online and as indicated below:

LastName, A.A.; LastName, B.B.; LastName, C.C. Article Title. *Journal Name* **Year**, *Article Number*, Page Range.

ISBN 978-3-03897-958-6 (Pbk)
ISBN 978-3-03897-959-3 (PDF)

Cover image courtesy of Carl F. Kusche and Rickmer Meya.

Contents

About the Special Issue Editor

Talal Al-Samman is a senior scientist and chief engineer at the Institute of Physical Metallurgy and Metal Physics of the RWTH Aachen University, where he also earned his Ph.D. in Engineering Sciences (2008). His research is concerned with advanced structural materials for lightweight applications, focusing on the science and engineering of magnesium alloys and their potential use to lighten automotive structures given their excellent strength-to-weight ratios. Over the past few years, he and his group have made influential contributions to understanding the deformation mechanisms, recrystallization and grain growth annealing phenomena, and crystallographic texture evolution in HCP metals, which dictate their mechanical performance and operational stability. On the basis of a comprehensive physical understanding of the underlying mechanisms of microstructure and texture evolution, he has introduced various concepts of microstructure and texture engineering in order to improve the cold formability and elevated temperature strength of commercial magnesium alloys. Given its solubility in the physiological environment of the human body, magnesium is currently considered to be a very promising biodegradable material for orthopedic implants and vascular stents. Dr. Al-Samman and his co-workers have also been active in this field, trying to address the biocorrosion and biodegradation behavior of newly developed ultralight Mg–Li–Ca alloys for potential degradable implant applications. He has published over 50 papers on these topics in peer-reviewed journals, and his research work is well recognized as internationally leading, as demonstrated by his numerous invitations for conference presentations, memberships in international conference advisory panels, and invitations from internationally renowned materials science journals to author featured articles.

Preface to "Material and Process Design for Lightweight Structures"

The use of lightweight structures across several industries has become inevitable in today's world given the ever-rising demand for improved fuel economy and resource efficiency. In the automotive industry, composites, reinforced plastics, and lightweight materials, such as aluminum and magnesium are being adopted by many OEMs at increasing rates to reduce vehicle mass and develop efficient, new lightweight designs. Automotive weight reduction with high-strength steel is also witnessing major ongoing efforts to design novel, damage-controlled forming processes for a new generation of efficient, lightweight steel components. Although great progress has been made over the past decades in understanding the thermomechanical behavior of these materials, their extensive use as lightweight solutions is still limited due to numerous challenges that play a key role in cost competitiveness. Hence, significant research efforts are still required to fully understand the anisotropic material behavior, failure mechanisms, and, most importantly, the interplay between industrial processing, microstructure development, and the resulting properties.

This Special Issue reprint book features concise reports on the current status in the field. The topics discussed herein include areas of manufacturing and processing technologies of materials for lightweight applications, innovative microstructure and process design concepts, and advanced characterization techniques combined with modeling of material's behavior. I hope that this collection of research will be of interest to a broad readership of material scientists, engineers, and students. I also hope that the broad scope of contributions and variety of topics presented will inspire researchers worldwide to conduct new cutting-edge studies that will lead to new discoveries and further enhancement in the field of lightweight, high-strength engineering structures.

Talal Al-Samman
Special Issue Editor

metals

MDPI

Editorial

Material and Process Design for Lightweight Structures

Talal Al-Samman [ORCID]

Institute of physical metallurgy and metal physics, RWTH Aachen University, Kopernikusstr. 14, 52074 Aachen, Germany; alsamman@imm.rwth-aachen.de

Received: 29 March 2019; Accepted: 4 April 2019; Published: 6 April 2019

1. Background

The ever-rising demand for increased fuel efficiency and a reduction in the harmful emission of greenhouse gases associated with energy generation and transportation has led, in recent years, to a resurgence of interest in light materials and new lightweight design strategies. In the automotive industry, the need to reduce vehicle weight has given rise to extensive research efforts to develop aluminum and magnesium alloys for structural car body parts. In aerospace, the move toward composite airframe structures urged an increased use of formable titanium alloys. In steel research, there are also major ongoing efforts to design novel damage-controlled forming processes for a new generation of efficient and reliable lightweight steel components. All of these materials, and more, constitute today's research mission for lightweight structures. They provide a fertile materials science research field aiming to achieve a better understanding of the interplay between industrial processing, microstructure development, and the resulting material properties.

Given the extensive scientific and technological importance of this timely subject, this Special Issue on "Material and Process Design for Lightweight Structures" was dedicated to collect concise reports on the current status in the field. The topics discussed herein include areas of manufacturing and processing technologies of materials for lightweight applications, innovative microstructure and process design concepts, advanced characterization techniques combined with modeling of materials behavior.

2. Contributions

The first article [1] is concerned with quantifying and minimizing damage in the complex microstructure of dual-phase steels. The study developed from a close and fruitful collaboration between the RWTH Aachen University and the Technical University of Dortmund within the context of the Collaborative Research Centre CRC/Transregio 188 "Damage-Controlled forming processes" supported by the German Research Foundation. The authors report outstanding technological advance on two fronts; In terms of (i) damage-reduced bending methodology (stress-superposed bending) and (ii) automated damage quantification of large micrographs of the order of a mm^2 at high resolution by means of panoramic imaging within a scanning electron microscope. A reduction of deformation-induced damage during processing of advanced high strength steels contributes greatly to lightweight design by allowing to offset material weight via the realization of thin-walled sheet metal parts while maintaining their mechanical performance.

The second of this set of articles is a featured paper by R. Kulagin et al. marking international research efforts between Germany, Ukraine and Australia in the area of severe plastic deformation (SPD) technologies and ultrafine-grained architectured materials [2]. The authors propose, for the first time, an analytical model capable of determining optimal deformation parameters and calculating the equivalent strain distribution over the entire sample length during the high pressure torsion extrusion process. This technique has the advantage over conventional high pressure torsion of being able to process long samples in a semi-continuous way, which is considered an exceedingly promising next step towards industrial upscaling of SPD. Owing to the simplicity and robustness of the presented theoretical approach, the authors reckon it can be successfully applied to calculating the mechanical behavior of lightweight structures made of magnesium, aluminum, and titanium.

In the third contribution [3], G. Cornacchia et al. present an innovative approach to further expand the field of application of high pressure die casting. This technology combines many advantages being able to produce thin-walled components, at low costs and high production volumes. The main limitation, however, is the difficulty to cast complex hollow components by the use of lost cores that are able to endure the high pressures used in the process. In this regard, the authors combine numerical and experimental research work to develop and utilize new ceramic cores that allow the production of an improved aluminum crossbeam for passenger cars. The study shows a promising avenue to implement this new technology for other safety relevant automotive hollow component.

B. Zhu and co-workers [4] provide an important insight of lightweight design into another field of applications, namely electrical transmission fittings, which are conventionally manufactured from cast or forged steel as heavyweight thick parts. The work demonstrates a new multilayered-sheet hot stamping process used to produce an electric-power-fitting product. The key challenge was to determine the optimum combination of the number of sheet layers and the contact pressure because of their significant effect on the final microstructure and mechanical properties. From numerous performance tests, a successful approach of using double-layered sheets was derived, achieving a fully martensitic microstructure at a relatively low contact pressure. The fabricated new component met the required standards for mechanical properties and load capacity, and exhibited a fantastic weight reduction of 60%.

The fifth contribution by S. Yi and co-workers from Helmholtz-Zentrum Geesthacht and Korea Institute of Materials Science is concerned with aspects of texture and microstructure control of a new non-flammable Mg-Al-Zn-Y-Ca (AZXW3100) magnesium sheet alloy, with the main goal of enhancing the ductility and formability at ambient temperatures [5]. The authors conducted a systematic investigation of the effect of the rolling temperature and imposed deformation per pass on weakening the basal texture and refining the microstructure during subsequent annealing treatments. Their findings clearly suggest that a rolling temperature of 450 °C and an increasing strain per pass (from 0.1 to 0.3) up to 11 passes, combined with following short 400 °C annealing can deliver a highly ductile and formable sheet exhibiting a remarkable Erichsen index of 8.1. Such huge success has significant implications for sheet metal forming perspectives, particularly in the automotive and aircraft industries, where modern, competitive magnesium alloys, due to their excellent strength-to-weight ratio, are becoming increasingly popular.

In "Effect of Surface Roughness on the Bonding Strength and Spring-Back of a CFRP/CR980 Hybrid Composite" [6], J.-H. Hwang et al. cover a critical subject in the field of lightweight hybrid composite materials, where they discuss possible improvements in the interfacial bonding behavior between the CFRP and the metallic material, and the springback response after forming by means of V-bending tests. Surface treatment experiments and lap shear adhesion tests were conducted to investigate the change in the bonding strength as a function of surface roughness, bonding pressure, compressive force, and compression direction. The results show a visible trend of increasing bonding strength with increasing surface roughness up to an optimal value, after which the occurrence of voids cause a fatal decrease in bond strength.

With a view to design connected processing steps that ensure viable manufacturing of lightweight components at high speeds, Rao et al. employed processing maps to investigate the hot working behavior of a new creep-resistant Mg–4Al–2Ba–2Ca alloy [7]. They report that in the as-cast condition, the alloy has a limited workability due to the presence of a large volume of intermetallic phases at the grain boundaries. To solve this problem, they introduced a connected step of extrusion, which helped greatly in refining the grain size and the particle distribution. They nicely show that the processing map for the extruded alloy exhibits a reduced flow instability regime, and a much more attractive workability window, characterized by suitable working temperatures to achieve a fine grain size, and sufficiently high strain rates to enable manufacturing at viable speeds.

The article by X. Xue et al. [8] fills another knowledge gap related to the sophisticated production of hollow, thin-walled aluminum alloy profiles, used for example in the bodies of high speed trains. The challenging aspect of extruding such profiles lies in their complex cross sections, which renders material flow in the extrusion die cavity much complex and difficult to control. As a result, the extruded parts are often liable to major defects leading to twisted or distorted profiles. Via numerical simulations and validation experiments, the authors present an optimum design of a die structure used in a multi-output porthole extrusion process, developed to reduce the forming load and improve the product quality. The current research provides a useful direction for obtaining a balanced metal flow behavior with uniform extrusion velocity that leads to minimizing extrusion defects of complex aluminum profiles during porthole die extrusion.

The important topic of fatigue and fracture behavior was also covered in this special issue by L. Zhan et al. in "Effect of Process Parameters on Fatigue and Fracture Behavior of Al-Cu-Mg Alloy after Creep Aging" [9]. The aim of the study was to analyze the effects of different creep aging parameters on the creep behavior, mechanical properties, and fatigue fracture behavior of a widely used Al-Cu-Mg alloy in the aerospace industry in order to advance the development of creep aging treatments of this class of aluminum alloys. The findings suggest that an increase of temperature and stress improves the creep response and fatigue life of the alloy up to a certain extent, which is then followed by a deterioration of these properties if the temperature and stress continue to increase. With the help of transmission electron microscopy, the authors conclude that the transition in properties is due to modified precipitation characteristics, and provide a clear concept on how to tune the microstructure to achieve optimal creep aging performance.

Finally, in the last article [10], K. Zheng et al. address the performance of in-die quenching during hot stamping of AA6082 aluminum alloy by means of systematic experimental and analytical investigations. The conducted work marks out numerous influencing factors, such as the initial work-piece and die temperatures, quenching pressures, work-piece thickness, and die clearances. The results revealed that the in-die quenching efficiency can be significantly enhanced by decreasing the initial work-piece and die temperatures. The authors also note that die clearances need to be carefully designed in order to obtain sufficiently high quenching rates and satisfying strength of hot-stamped panel components. The study provides useful, practical insights into designing manufacturing processes of hot stamping parts for mass production.

3. Concluding Remarks

As a Guest Editor of this special issue of Metals I greatly enjoyed reading and learning so much from the above mentioned articles. The broad scope of contributions and accomplishments is truly remarkable, and emphasizes the wide variety of topics that could and should be treated under this rapidly-growing and far-reaching subject. I hope that with this special issue we were able to provide the readers with a sense of where significant advances are being made, where critical issues remain pending, and, from the authors' perspectives, where the field is heading in the near future.

Acknowledgments: The guest editor would like to express his deepest appreciation to all the authors who contributed their research to this special issue. I also want to thank all the reviewers for their extremely valuable and timely feedback that allowed all the authors to push the quality of their contributions to a higher level. Special thanks to Ms. Betty Jin and the editorial staff for their enduring efforts in bringing this special issue together. Last but not the least, the Guest Editor is very grateful for continual financial support from the Deutsche Forschungsgemeinschaft (DFG) through numerous research grants within the field of lightweight material and process design.

Conflicts of Interest: The author declares no conflict of interest.

Metals **2019**, *9*, 415

References

1. Meya, R.; Kusche, C.F.; Löbbe, C.; Al-Samman, T.; Korte-Kerzel, S.; Tekkaya, A.E. Global and High-Resolution Damage Quantification in Dual-Phase Steel Bending Samples with Varying Stress States. *Metals* **2019**, *9*, 319. [CrossRef]
2. Kulagin, R.; Beygelzimer, Y.; Estrin, Y.; Ivanisenko, Y.; Baretzky, B.; Hahn, H. A Mathematical Model of Deformation under High Pressure Torsion Extrusion. *Metals* **2019**, *9*, 306. [CrossRef]
3. Cornacchia, G.; Dioni, D.; Faccoli, M.; Gislon, C.; Solazzi, L.; Panvini, A.; Cecchel, S. Experimental and Numerical Study of an Automotive Component Produced with Innovative Ceramic Core in High Pressure Die Casting (HPDC). *Metals* **2019**, *9*, 217. [CrossRef]
4. Zhu, B.; Zhu, Z.; Jin, Y.; Wang, K.; Wang, Y.; Zhang, Y. Multilayered-Sheet Hot Stamping and Application in Electric-Power-Fitting Products. *Metals* **2019**, *9*, 215. [CrossRef]
5. Yi, S.; Victoria-Hernández, J.; Kim, Y.; Letzig, D.; You, B. Modification of Microstructure and Texture in Highly Non-Flammable Mg-Al-Zn-Y-Ca Alloy Sheets by Controlled Thermomechanical Processes. *Metals* **2019**, *9*, 181. [CrossRef]
6. Hwang, J.; Jin, C.; Lee, M.; Choi, S.; Kang, C. Effect of Surface Roughness on the Bonding Strength and Spring-Back of a CFRP/CR980 Hybrid Composite. *Metals* **2018**, *8*, 716. [CrossRef]
7. Rao, K.; Chalasani, D.; Suresh, K.; Prasad, Y.; Dieringa, H.; Hort, N. Connected Process Design for Hot Working of a Creep-Resistant Mg–4Al–2Ba–2Ca Alloy (ABaX422). *Metals* **2018**, *8*, 463. [CrossRef]
8. Xue, X.; Vincze, G.; Pereira, A.; Pan, J.; Liao, J. Assessment of Metal Flow Balance in Multi-Output Porthole Hot Extrusion of AA6060 Thin-Walled Profile. *Metals* **2018**, *8*, 462. [CrossRef]
9. Zhan, L.; Wu, X.; Wang, X.; Yang, Y.; Liu, G.; Xu, Y. Effect of Process Parameters on Fatigue and Fracture Behavior of Al-Cu-Mg Alloy after Creep Aging. *Metals* **2018**, *8*, 298. [CrossRef]
10. Zheng, K.; Lee, J.; Xiao, W.; Wang, B.; Lin, J. Experimental Investigations of the In-Die Quenching Efficiency and Die Surface Temperature of Hot Stamping Aluminium Alloys. *Metals* **2018**, *8*, 231. [CrossRef]

metals MDPI

Article

Global and High-Resolution Damage Quantification in Dual-Phase Steel Bending Samples with Varying Stress States

Rickmer Meya [1,*], Carl F. Kusche [2], Christian Löbbe [1], Talal Al-Samman [2], Sandra Korte-Kerzel [2] and A. Erman Tekkaya [1]

[1] Institute of forming technology and lightweight components, TU Dortmund University, Baroper Str. 303, 44227 Dortmund, Germany; christian.loebbe@iul.tu-dortmund.de (C.L.); erman.tekkaya@iul.tu-dortmund.de (A.E.T.)
[2] Institute of physical metallurgy and metal physics, RWTH Aachen, Kopernikusstr. 14, 52056 Aachen, Germany; kusche@imm.rwth-aachen.de (C.F.K.); alsamman@imm.rwth-aachen.de (T.A.-S.); korte-kerzel@imm.rwth-aachen.de (S.K.-K.)
* Correspondence: rickmer.meya@iul.tu-dortmund.de; Tel.: +49-231-755-2669

Received: 31 January 2019; Accepted: 6 March 2019; Published: 12 March 2019

Abstract: In a variety of modern, multi-phase steels, damage evolves during plastic deformation in the form of the nucleation, growth and coalescence of voids in the microstructure. These microscopic sites play a vital role in the evolution of the materials' mechanical properties, and therefore the later performance of bent products, even without having yet led to macroscopic cracking. However, the characterization and quantification of these diminutive sites is complex and time-consuming, especially when areas large enough to be statistically relevant for a complete bent product are considered. Here, we propose two possible solutions to this problem: an advanced, SEM-based method for high-resolution, large-area imaging, and an integral approach for calculating the overall void volume fraction by means of density measurement. These are applied for two bending processes, conventional air bending and radial stress superposed bending (RSS bending), to investigate and compare the strain- and stress-state dependent void evolution. RSS bending reduces the stress triaxiality during forming, which is found to diminish the overall formation of damage sites and their growth by the complimentary characterization approaches of high-resolution SEM and global density measurements.

Keywords: damage; characterization; automated void recognition; density; bending; stress superposition

1. Introduction

Over the past years, processes of damage formation have yielded tremendous interest in the field of materials science, due to the rising demand for advanced metallic materials combining high strength and excellent formability. For many of those materials, damage formation is a point that has to be addressed due to their intrinsic microstructural heterogeneity [1]. Typically, damage formation and accumulation take place during plastic deformation and are most commonly observed as the formation and growth of voids [2]. The interaction of these voids ultimately leads to failure; however, the mechanisms of damage formation and evolution themselves are not part of the process of material failure. During plastic deformation, processes of void nucleation, evolution and coalescence take place and lead to a continuous degradation of mechanical properties, and ultimately, failure.

Before the interaction and coalescence of voids start, void growth is the main mechanism of damage evolution. This process has been extensively researched, especially in the field of modeling, ranging from the fundamental modelling of void growth [3] and also nucleation [4] to advanced, high-resolution microstructural simulations [5]. As experimental approaches as well as the modelling of void growth have shown, the growth behavior of microstructural voids is largely dependent not only on the magnitude of strain, but in a significant way on the applied stress state [6].

For structural parts in the automotive industry, high-strength values combined with good formability are required; this objective has, recently, mainly been achieved by the usage of advanced high-strength steels (AHSS). A widely used variety of this class are the dual-phase steels. These combine low production costs compared to other AHSS with beneficial ductility, high yield strength values and near-linear strain hardening properties [7]. These properties are realized by a microstructure made up of ferritic and martensitic constituents. However, the complementary properties of these constituents typically cause a strong contrast in plastic deformation between the two phases, leading to a stress and strain partitioning behavior in the local microstructure. This incompatibility leads to the nucleation of voids, caused by distinct mechanisms [8]; the hard martensite islands are prone to locally brittle damage initiation, i.e. martensite cracking. These cracks typically occur at prior austenite grain boundary sites [9]. In addition to this mechanism, decohesion processes at interfaces such as phase boundaries between martensite and ferrite or at ferrite grain boundaries can take place [10]. In many cases, the local morphology [11] and heterogeneity of the microstructure [12] is the main factor determining the dominant damage mechanism, and a wide variety of intermediate forms or combinations of the above-mentioned mechanisms are observed. Commercially used dual-phase steels such as the one employed in this work often show a significant banding of martensite, leading to a pattern of voids often described in the literature as "necklaces" [12]. These agglomerations of voids, typically observed at large strains, are caused by the basic mechanisms of martensite cracking, phase boundary and grain boundary decohesion, but represent a distinct pattern of damage sites in their own right.

In order to link damage formation and stress state, independent parameters—namely the Lode angle parameter, $\bar{\theta}$, and the stress triaxiality η—are used. Both parameters influence the damage evolution [13]. The stress triaxiality, η, is defined as the ratio of hydrostatic stress, σ_h, and the von Mises equivalent stress, σ_{vM}:

$$\eta = \frac{\sigma_h}{\sigma_{vM}} \tag{1}$$

The hydrostatic stress is thought to be responsible for the growth, or if negative, even shrinking of already nucleated voids in the microstructure. It is therefore expected for stress states with lower stress triaxialities to cause a delayed void evolution for forming-induced damage. With the deviatoric stress tensor, σ^{dev}, the third normalized invariant, ξ, can be derived:

$$\xi = \frac{27 \det\left(\sigma^{dev}\right)}{2\,\sigma_{vM}^3} = \frac{27/2 \cdot (\sigma_1 - \sigma_h) \cdot (\sigma_2 - \sigma_h) \cdot (\sigma_3 - \sigma_h)}{\left\{ \left[(\sigma_1 - \sigma_2)^2 + (\sigma_2 - \sigma_3)^2 + (\sigma_3 - \sigma_1)^2 \right]/2 \right\}^{2/3}} \tag{2}$$

This invariant ξ is defined in the range of $-1 \leq \xi \leq 1$. The normalized Lode angle parameter, $\bar{\theta}$, is defined as

$$\bar{\theta} = 1 - \frac{2}{\pi} \arccos(\xi) \tag{3}$$

During plane strain plastic forming, the second principal stress is always

$$\sigma_2 = \frac{\sigma_1 + \sigma_3}{2} \tag{4}$$

In the bending of sheet with a much larger width compared to the thickness, plane strain deformation conditions can be assumed. This leads to a constant normalized Lode angle parameter, $\bar{\theta} = 0$.

Anderson et al. [14] revealed that the strain to fracture for a lower triaxiality is lower compared to higher triaxialities for a constant Lode angle parameter in DP800 steels. Thus, the stress state is important for material failure, but it also influences damage evolution, as failure can be the consequence of damage. Technologically, the stress state during bending must then be influenced to reduce damage. Technological solutions are, for instance, bending with a solid counter punch [15], roll bending with additional rolls [16], bending with an elastomer [17] and radial stress superposed bending [18]. Bending using elastomers is capable of reducing the stress triaxiality during bending by applying a counter pressure due to the inserted elastomer. This leads to a delayed damage evolution in terms of void nucleation, which subsequently influences the fatigue lifetime of bent products [19]. Thus, the accumulation of damage during forming is important for lightweight design and has to be taken into account as it affects the product performance. For industrial purposes, elastomer-bending is not feasible for controlling the stress state, as the elastomer does not apply reproducible counter pressures during forming and is limited in the magnitude of applicable stresses (the maximum pressure is less than 150 MPa), as well as showing a rapid degradation over its lifetime. Recently, a new bending process with predetermined stress states was introduced [20]. The so-called radial stress superposed bending (RSS bending) is capable of reducing the stress triaxiality and applying pressures up to the flow stress of the material in a reproducible way. It has already been shown to protract damage nucleation, leading to a reduced number of voids [20].

For the product design or process modeling, the amount of damage can be expressed directly as the area or volume fractions of voids or indirectly via certain mechanical properties. Lemaitre and Dufailly (1987) showed eight methods for direct and indirect damage measurement techniques and rated their suitability [21]. Direct measurements include microscopic analysis, X-ray analysis and density measurements. Indirect damage measurements are, for example, the decrease in Young's modulus, micro hardness or indentation modulus [22]. For damage quantification, direct measurements are preferable as there is no mathematical model connected to the calculation of damage quantity. A damage variable, D_s, in surface observations is proposed by Lemaitre and Dufailly as the ratio of the void area, S_d, and the undamaged area, S [21].

For a DP600, the void volume fraction before failure is usually below 1–2% of the whole volume [22]. Consequently, the preparation of specimens for direct surface measurements is challenging. Samuels et al. showed that mechanical polishing might introduce strain hardening in the material surface [23]. Also, a void smearing effect could be shown due to different polishing steps [24]. Isik et al. revealed that ion beam slope cutting is capable of analyzing void sizes down to 0.05 μm^2 [25]. Another quantification method is radiography. Using X-ray microtomography, specimens can be analyzed without metallographic preparation in a non-destructive way; this method is, however, limited by its spatial resolution [26].

For an integral approach to measuring void volume fractions, density measurements can be applied. Ratcliffe presented a method for measuring small density changes in solids using the Archimedean principle [27]. Schmitt et al. showed that different strain paths lead to different relative density changes [28]. Bompard proved the possibility of measuring density changes in a tensile specimen and correlated this to damage [29]. The method has equally been applied by Lemaitre and Dufailly to quantify damage evolution [21].

Lapovok et al. measured the density of specimens in a continuously cast aluminum alloy formed in an equal channel angular drawing process with the help of the Archimedean principle [30]. Tetrachloroethylene with a density of 1.62 g/cm^3 was used instead of distilled water for higher accuracy. They correlated the change in density to the stress and strain state that is responsible for different paths of damage evolution. Tasan et al. stated that tactile density measurements are not capable of analyzing damage for specimens with a volume of as low as 1 mm^3 for spatially resolved measurements [22] as the scatter observed for small volumes dominates the measurements.

Thus, in the current state of the art, it is shown that stress superposition during bending leads to delayed fracture. Despite this, it is not clear what influence the lowered stress triaxiality has on the void evolution and damage mechanisms. To quantify and characterize damage in bent samples, the methods for automated void recognition and density measurements have to be adopted to the requirements set by bending samples. With these characterization tools, the influence of the alteration in stress state on damage evolution can be quantified and subsequently used for the modelling or prediction of the expected service life time.

2. Materials and Methods

The DP steel applied in this study is of DP800 grade, which usually indicates that it has a guaranteed tensile strength of more than 800 MPa and its microstructure consists mainly of ferritic and martensitic constituents. However, a very small fraction of remaining austenite and bainite might still be present in the microstructure in small volume percentages. The as-received DP800 sheet material was subjected to a hot-dip galvanizing process using a zinc bath, which provides the rolled sheets with corrosion protection. The average grain size ranges from 2 µm to 20 µm, with martensite particles of approx. 2 µm in diameter embedded in the matrix. The characterized microstructure material shows a strong banding of the martensite phase along the rolling direction (Figure 1).

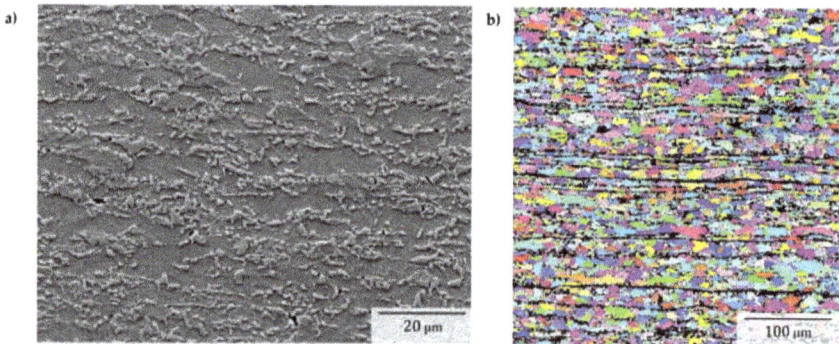

Figure 1. (**a**) Microstructure of the used dual-phase DP800 steel imaged by SEM, with visible deformation-induced voids. (**b**) Electron-backscatter-diffraction mapping of ferrite grains; martensite bands are visible as black areas.

The flow curve at room temperature (obtained by a Zwick Z250 universal testing machine, ZwickRoell GmbH & Co. KG, Ulm, Germany) is given by experimental data from uniaxial tensile tests and extrapolated according to Gosh (Figure 2).

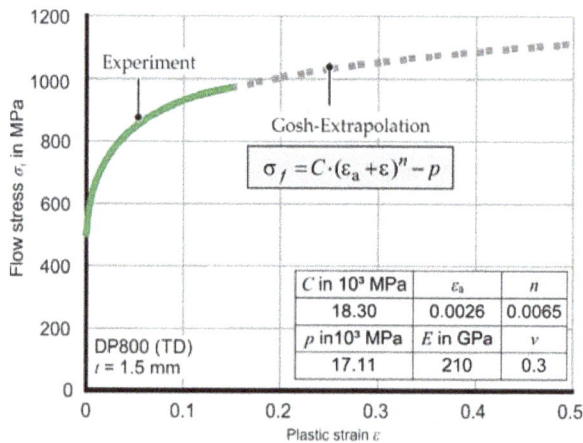

$$\sigma_f = C \cdot (\varepsilon_a + \varepsilon)^n - p$$

C in 10^3 MPa	ε_a	n
18.30	0.0026	0.0065

p in 10^3 MPa	E in GPa	v
17.11	210	0.3

Figure 2. Flow curve of the investigated DP800 steel with experimental data and extrapolation according to Gosh.

The tensile tests were conducted with a specimen geometry (DIN 50125—H 20 × 80) according to DIN EN ISO 6892-1 with a velocity of 0.0067 s^{-1} to ensure a constant strain rate. The measurement of the elongation was done directly on the test sample with a tactile macro-extensometer (Gauge length of 80 mm, ZwickRoell GmbH & Co. KG, Ulm, Germany). The flow curve is derived up to the uniform elongation experimentally and then extrapolated. The extrapolation parameters (ε_a: strain at yielding, n: hardening exponent, C and p: fitting parameters) according to Gosh are derived with the least square fitting method. The Young's modulus E and Poisson's ratio v are given in Figure 2.

2.1. SEM Panoramic Imaging, Void Recognition and Area Determination

Deformation-induced damage in these grades of dual-phase steels typically occurs in the form of microscopic voids with sizes in the range of several hundred nm [8] to a few μm [26]. To reliably quantify voids at such small scales, high-resolution measurements of large micrographs in the order of mm^2 are required. This was achieved in the present work by employing advanced scanning electron microscopy (LEO 1530, Carl Zeiss Microscopy GmbH, Jena, Germany) combined with panoramic imaging and an image stitching algorithm based on the VLFeat Matlab toolbox [31]. All panoramic images have been obtained at the tip of the bending sample (Figure 3) at a resolution of 32 nm/px using secondary electrons (SE) and a 20% area overlap. The field width of a single image was 100 μm, resulting in a total panoramic image size of 1000 μm × 500 μm. Respective specimens were mechanically polished to 0.25 μm and subsequently etched in 1% Nital for 10 s. A consistent, light etching is critical for this method, as shadowing effects of the protruding martensite phase have to be minimized for a reliable automated image recognition. The panoramic images are subsequently split into 5 slices that follow a radial direction. This approach is chosen to ensure an accurate measurement of the respective distance to the outer radius, which would be altered as, in bending samples, the upper edge cannot be straight. A binning of 3000 pixels in a radial direction was applied, and each data point was attributed to the middle of this bin, resulting in the outermost value for the distance to the outer radius being calculated as 48 μm from the edge.

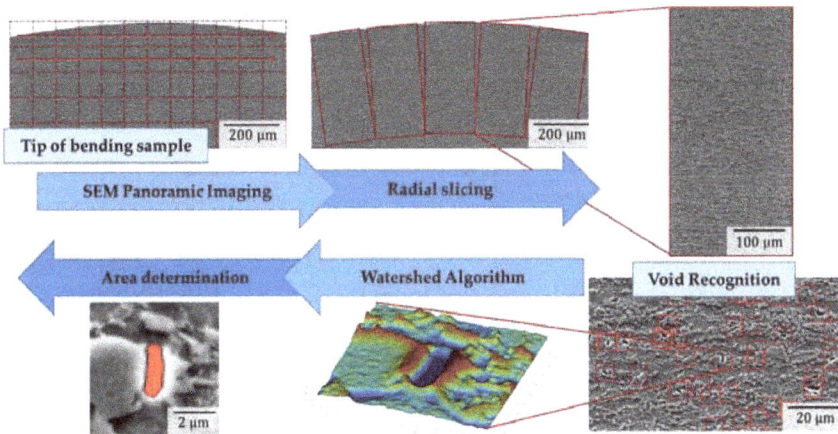

Figure 3. Schematic representation of void area measurements from panoramic SEM imaging. Individual images are stitched and voids recognized via a grayscale threshold. The identified voids are then individually processed using a watershed algorithm to measure their size.

Voids are identified using a grayscale threshold and the located sites from the original image processed further by the use of a watershed algorithm [32]. Here, by altering the grayscale threshold around the individual void, the optimum value for measuring the entire void area, but none of the surrounding microstructure, is determined. The calculation of the pixel areas results in a separate measurement for each individual void. This approach makes it possible to collect data not only for overall void area fractions, but for any type of measurement where information about each individual damage site is required. To calculate area fractions over one spatial coordinate, a moving bin is applied to smooth out peaks in the area fraction generated by single, larger voids, made possible by the individual identification and localization of damage sites.

A considerable error in the void area measurements of deformation-induced damage voids is generated by inclusions. These, in commercial DP steel, typically being TiN, can be caused to fall out of the polished surface during preparation, leaving voids of a similar, slightly larger diameter in the observed images. Examples of this type of voids as well as the above-mentioned fundamental mechanisms of damage nucleation and formation are shown in Figure 4. These voids at the sites of inclusions have a very different morphology from martensite cracks or interface decohesion sites, making them an ideal subject for recognition by deep learning [33]. In this work, neural networks have been trained using an initial data set, with the goal of automatically detecting voids that have been caused by inclusions in the microstructure. A system for recognizing these inclusions from SEM pictures has been developed and tested to an accuracy of over 95% and is applicable to the SEM panoramas used in this study [33]; here, it is used for inclusion void recognition only. With smaller plastic equivalent strains towards the sheet center, voids caused by inclusions become increasingly dominant, as this type of observed void is in the majority of cases an artefact of metallographic preparation by mechanical polishing. Non-metallic inclusions leave the surface during this process, and therefore cause surface voids that did not develop during plastic deformation and are therefore not to be measured simultaneously with plasticity-induced damage. The error normally introduced by measuring these inclusions as part of the void fraction is avoided by the recognition and exclusion of these particular voids.

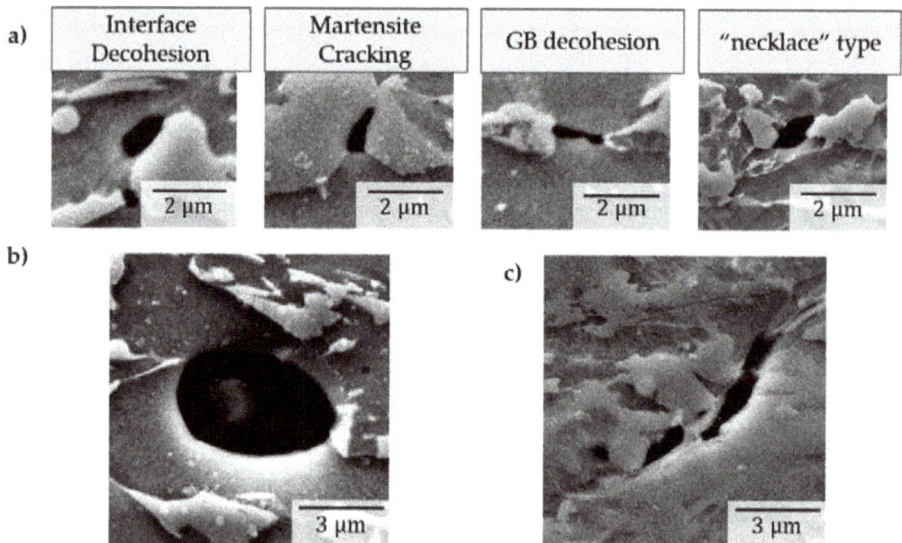

Figure 4. (**a**) Examples of fundamental damage mechanisms of martensite/ferrite interface decohesion, martensite cracking, ferrite grain boundary (GB) decohesion and "necklace" type voids between two adjacent martensite particles; (**b**) void observed due to (partly) removed inclusions from the surface, recognized by a deep learning algorithm; (**c**) void coalescence near the outer surface of an air bending sample.

The panoramic imaging method was carried out in the *x–y* plane ("in-plane") of a bending sample. To analyze all spatial directions, a second measurement was taken on a plane in the middle of the bending zone, parallel to the bending axis in *z* direction ("cross-section") (Figure 5).

a)

b)

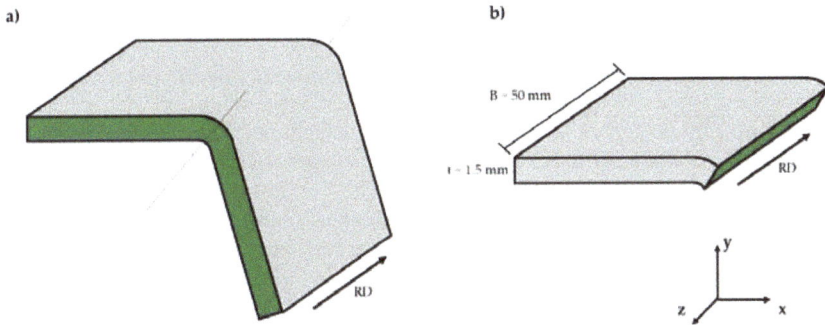

Figure 5. Evaluated planes in the bending samples (illustration): (**a**) cutting plane parallel to the bending plane ("in-plane", *x–y* plane); (**b**) cutting plane parallel to the bending axis ("cross-section", *y–z* plane). RD represents the rolling direction of the sheet.

The results were taken from these two perpendicular planes as the evolution and growth of voids are expected to be highly anisotropic due to the tensile stresses perpendicular to the bending radius, which will affect the morphology of voids.

2.2. Density Measurement Method

Density measurements of heavy metals for damage quantification require a high resolution in the order of 0.002 g/cm³ as void volume fractions as low as 0.2% are investigated. The measurement principle is based on hydrostatic weighing. A solid immersed in liquid apparently reduces its weight by the liquid volume weight. It is necessary to know the density of the liquid to measure the density of the submerged solid. In contrast to the classical Archimedean density measurement, the volume of the displaced liquid is not measured by an overflow, but the weight differences are measured. Therefore, this method is strongly dependent on the volume of the specimen, as higher volumina lead to higher precision. The IMETERV6 device (IMETER/MSB Breitwieser MessSysteme, Augsburg, Germany) is used for the measurements in this work. The measuring process consists of taring the specimen holder in the measuring liquid at the predefined immersion depth, then withdrawing and connecting to the specimen (Figure 6).

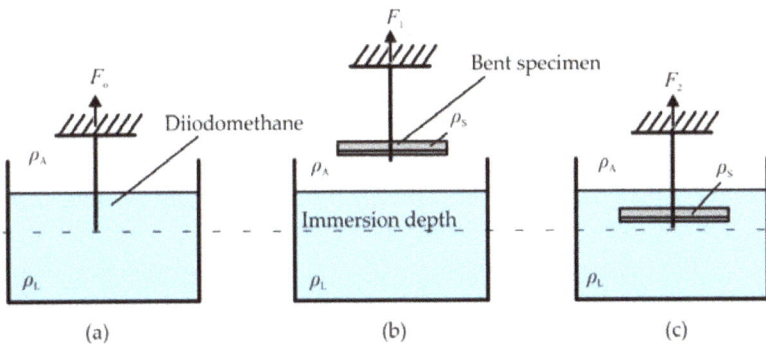

Figure 6. Buoyancy principle for density measurement: (**a**) calibration of the suspension in ambient air; (**b**) weighing of the specimen in ambient air; (**c**) weighing of the specimen in the measuring liquid.

Subsequently, the weighing of the specimen takes place in ambient air. Afterwards, specimens are immersed to exactly the same immersion depth with the same contact angle of the attached wire likewise at taring and weighing the buoyancy after the same predefined diving time. The same contact angle of the wire and liquid as well as the predefined immersion time are important for the measuring accuracy. Withdrawing and immersional weighing is repeated several times (15 to 25 times for 10 to 15 min) until the buoyancy values are constant. Thus, finally, all air bubbles from the surface are washed away and differences in the temperature of specimen and liquid have been compensated. The specimen is connected to a load cell (Sartorius AG, Göttingen, Germany) with a thin tungsten wire ($D = 40$ µm). The higher the density of the liquid, the higher the lifting force and therefore the measuring accuracy. Diiodmethane (CH_2I_2) is used as the immersion liquid, which is a so-called heavy liquid [34]. The measured density is 3.3027 g/cm^3 at 25 °C [35]. The density of a solid ρ_s can thus be determined according to physical relationships [35]. This calculation requires the density of the liquid ρ_L, the density of the ambient air ρ_A, the resulting force on the specimen outside the measuring liquid W_1 and that immersed in the measuring liquid, W_2.

$$\rho_s = \frac{\rho_L - \rho_A}{1 - \frac{W_2}{W_1}} + \rho_A \qquad (5)$$

To calculate the void volume fraction in the bent product, specimens are cut out of the bending zone and the bending leg of the samples. Since the properties of a bent part deviate at the edge of the sheet (the assumption of plane strain is invalid), only the constant area of the bending zone is examined (5 mm apart from the outer edges) (Figure 7).

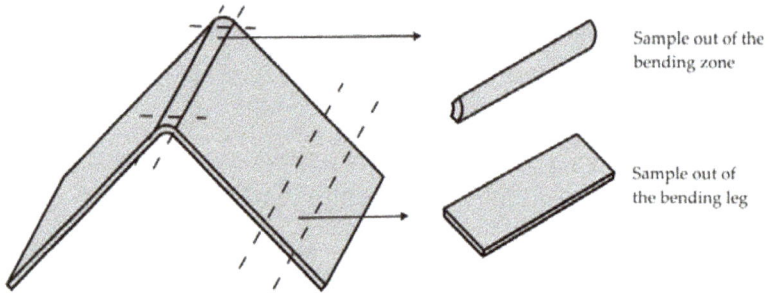

Sample out of the bending zone

Sample out of the bending leg

Figure 7. Area of sampling in a bent profile.

Comparative samples from the bending leg are examined. To further increase the measurement accuracy, three bent samples are simultaneously measured to increase the total volume to ~1 cm^3.

2.3. Air and Stress Superposed Bending Processes

The investigated air-bending process is defined by the geometrical parameters: die width w_d, punch radius r_p and die radius r_d (Figure 8). The sheet is laser cut (100 mm × 50 mm × 1.5 mm) and bent parallel to the rolling direction to an unloaded bending angle of 66°.

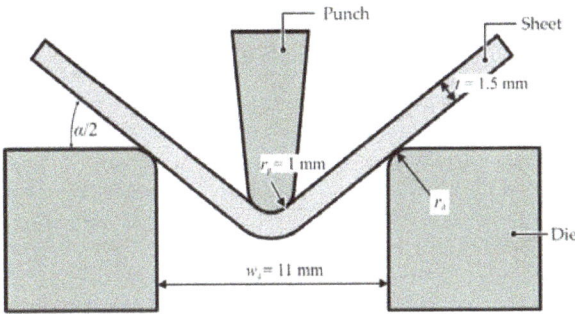

Figure 8. Air-bending process.

During air bending, the outer fibers are formed under a tensile stress state, whereas the inner fibers are contracted. The plastic strains increase towards the outer fiber. Therefore, forming-induced voids and the final failure occur at the outer fiber [36]. To delay the void evolution, radial stress superposed bending can be used [20]. In this process, defined compressive stresses are superposed during bending. A normal force N_r rotates around the outer bending fiber and superposes stresses. The technological implementation of the process is done by rotating tools in bearing shells driven by a hydraulic cylinder (Figure 9). The hydraulic cylinder is connected to the lower moveable bearing shell and is capable of applying a given, constant pressure.

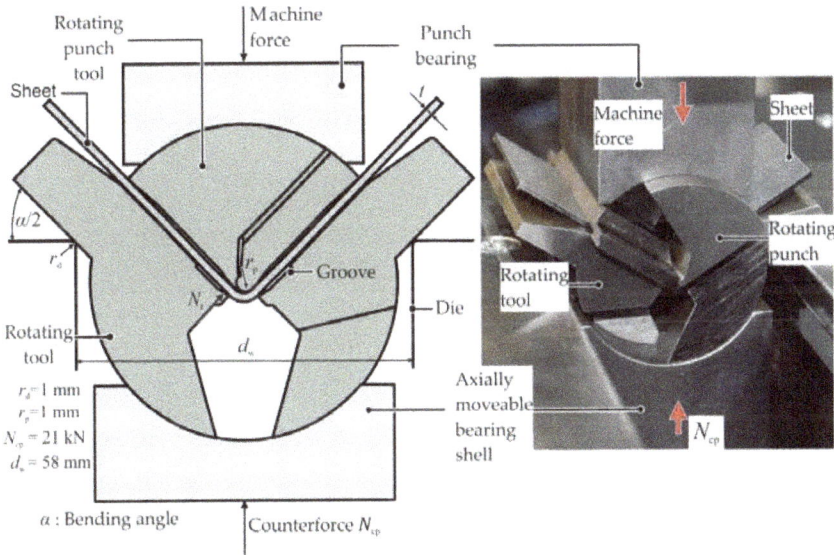

Figure 9. Technological implementation of radial stress superposed (RSS) bending.

The applied process parameters lead to a comparable geometry of the bent products. For the validation of the identical strain at the outer fiber, optical measurements and hardness measurements are used. Stress triaxiality values and the strain distribution over the sheet thickness is investigated numerically according to the model presented in previous work [37]. The FEM-modeling was carried out with the elastic-plastic modelling with ABAQUS2016/Implicit2D (Dassault Systèmes, Vélizy-Villacoublay Cedex, France). Plane strain conditions and a planar symmetry were assumed to reduce computational effort. The sheet is modelled with the flow curve given in Figure 2. The rotating tools are modelled to be purely elastic, and all other components are rigid. The smallest element size in the bending zone is 0.05 mm and in the bending leg is 0.3 mm. The friction between the tools is modelled by the Coulomb friction law ($\mu = 0.02$ between the lower rotating tools and the sheet/lower bearing shell, since it is lubricated; $\mu = 0.1$ between the upper rotating tool and the sheet/upper bearing shell; $\mu = 0.1$ in air bending). The maximum force deviation between the numerical and experimental punch force was lower than 10% [37].

Corresponding to Meya et al. [37], the stress triaxiality in air bending is $\eta_{min} = 0.57$, while the minimum stress triaxiality during RSS bending in this set-up is calculated as $\eta_{min} = -0.06$ at the outer fiber due to the superposed stresses.

3. Results

3.1. SEM-Based Damage Characterisation and Quantification

The measurements obtained by SEM observation yielded results for the global quantification of deformation-induced voids in the bent samples in the form of area fraction calculations, as well as achieving magnifications high enough to gain microstructural information about the individual mechanisms of damage nucleation and evolution, which have, however, not been considered in this work.

Firstly, the dominance of voids originating from inclusions in the steel microstructure becomes obvious when regarding the panoramic images; with greater distances from the outer radius, larger voids can almost exclusively be recognized as being caused by inclusions being removed from the polished surface during the preparation of the sample.

Secondly, regarding deformation-induced damage, both the absolute number of voids (Figure 10) and the mean size (Figure 11) of the voids are observed to increase towards the outer radius—the zone of the highest plastic equivalent strain. These observations are true for both applied bending methods and both observation planes. Representative examples of typical voids at the outer radius, at a distance of approx. 80 µm below the surface and at 400 µm below the outer radius, are given in Figure 11.

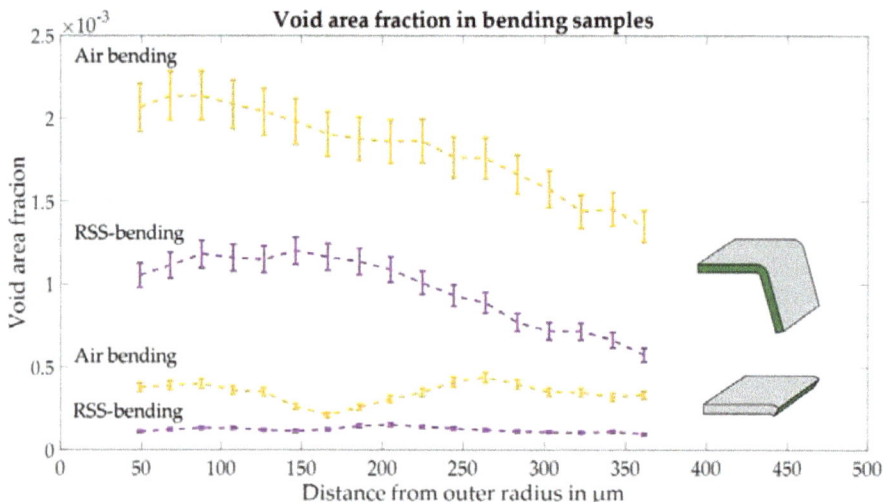

Figure 10. Void area fraction in air bending and RSS bending in the outer fibers measured by SEM imaging and automated void recognition.

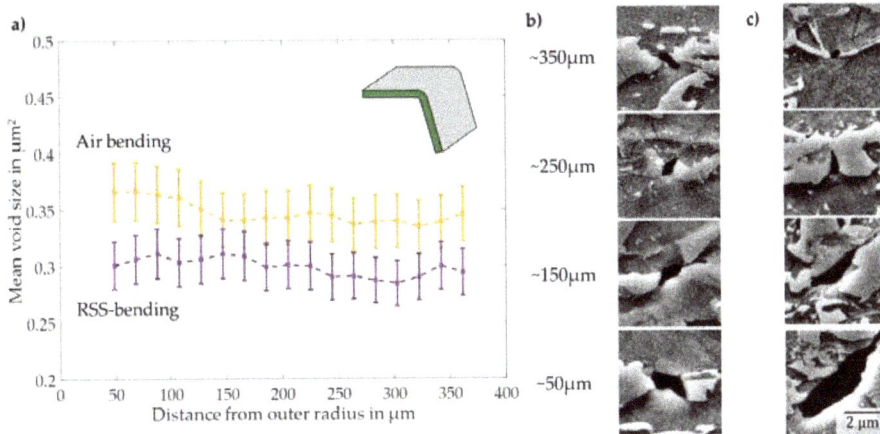

Figure 11. (a) Mean void size in the in-plane samples for air-bent and RSS-bent samples; representative void morphologies at various distances from the outer radius for (b) RSS bending and (c) air bending.

For each data point, an area of 33,379 μm^2 was evaluated. The error for the measurements is calculated via the error for the watershed algorithm value for each single void, and summarized over all voids in the field of observation.

Comparing the area fractions, voids per area ($n = 5345$ voids/mm^2 vs. $n = 3589$ voids/mm^2 at the outer fiber) and void sizes ($S = 0.37$ μm^2 vs. $S = 0.30$ μm^2 at the outer fiber) across the two applied processes, a significant change in all these quantities is observable for the radial stress superposed (RSS) bending method compared with conventional air-bending. As calculating the void area fraction for various regions of the sample takes into account the overall magnitude of voids existing and nucleating in the observed area and their sizes, this property is able to deliver a more complete picture of void nucleation and growth compared to solely regarding void numbers or mean sizes. Area fraction calculations from the detected voids show an average decrease of 52.1% for the RSS bending.

The void area fractions measured in the cross-section are significantly smaller (up to 4 times), as the imaging plane is perpendicular to the bending strain ε_x and no macroscopic strain in the z-direction occurs. However, in these images an average decrease in void area fractions up to 69% for the triaxiality-reduced RSS-bending process is also visible.

Measurements in as-received, undeformed samples only found 3 to 5 voids that could not be reliably attributed to being caused by inclusions on the same field of view as used in the measurements above. Therefore, the void area fractions for the undeformed state of the sheet metal are assumed to be non-existent.

The results for mean void areas equally show a lower average void area for the RSS-bending process; this effect is, however, largely dominated by the scatter in void sizes, in particular at the region near the outer radius. On the one hand, large voids are found in this region in the air-bending samples, whereas the RSS-bending process does not provoke these large void sizes. A typical evolution of voids for both bending processes is shown in Figure 11. However, a clear difference in size emerges when regarding maximum void sizes instead of mean values; while for the air-bending process, void sizes at the outer radius reach up to 2.26 μm^2, RSS-bent samples only showed void sizes lower than 1.22 μm^2. This tendency does not appear in the mean void size calculations, as these are dominated by the large number of nucleating voids below 0.3 μm^2 in size. This tendency for the growth of voids in air-bending samples can be underlined by normalizing the mean void sizes not by number, but by their respective area fractions. While for air bending, 74.9% of the total void area is made up by voids larger than the calculated mean void size per bin, this fraction calculates to a smaller value of 65.7% for the RSS-bending samples. Even though there is no clear threshold in size or morphology after which a void can clearly be classified as "grown", this statistical approach shows a difference in the composition of the cumulative void area from small and large void sizes.

3.2. Density Measurements

The specimen is not tempered in advance, so it needs time to adapt its temperature to the measuring fluid. Also, the tempering of the measuring cup fluctuates because of the specimen immersion. The measurements are repeated several times to compensate and stabilize the influences of temperature differences and air bubbles (Figure 12).

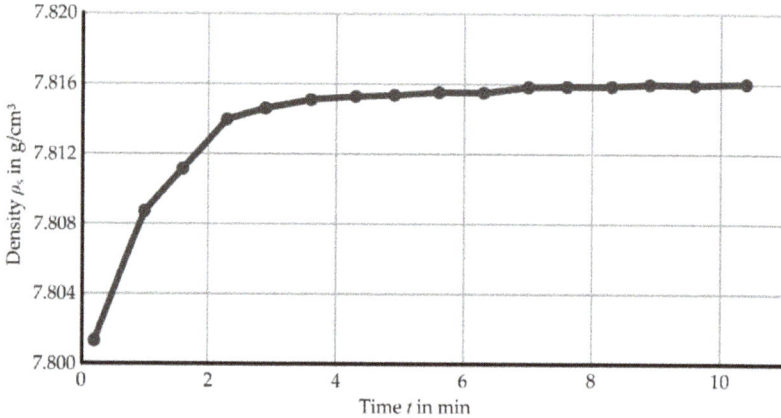

Figure 12. Density fluctuation due to bubbles and alternating temperature over time for an unbent sheet.

The overall uncertainty regarding the measured density is a function of the measuring accuracy of the loading cell, the tempering, the ambient air and fluid density as well as the calculated temperature dependency of the fluid. The uncertainty reduces with a higher specimen volume and higher density of the fluid. In this set-up, the average uncertainty of the density is 0.0021 g/cm^3 for a specimen volume of 1 cm^3. The influence of the altered stress triaxiality on macroscopic density and the uncertainty of different specimens shows a maximum fluctuation of $\pm$ 0.0008 g/cm^3 (Figure 13).

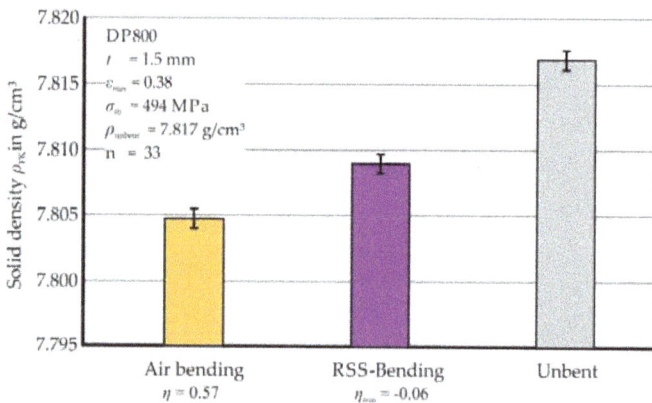

Figure 13. Density of different bent products in comparison to the as-received material.

The density of the air-bent part (η_{min} = 0.57) is reduced by 0.15% compared to the as-received material. In comparison, the RSS-bent (η_{min} = −0.06) product's density is reduced by 0.10%. Thus, a reduction of 33% in density loss is achieved by using the applied stress superposition during bending.

3.3. Resulting Strains and Stresses in Bending

The maximum curvature is measured by light optical microscopy and is revealed to be 0.307 ± 0.002 mm^{-1} for air bending and 0.310 ± 0.004 mm^{-1} for RSS-bending samples at the symmetry axis of the bending area. This difference of around 1% in curvature indicates similar plastic strains at the outer fiber. The numerically investigated strain distribution is also comparable, since the deviation of the equivalent plastic strain over the investigated area is less than 10% (Figure 14).

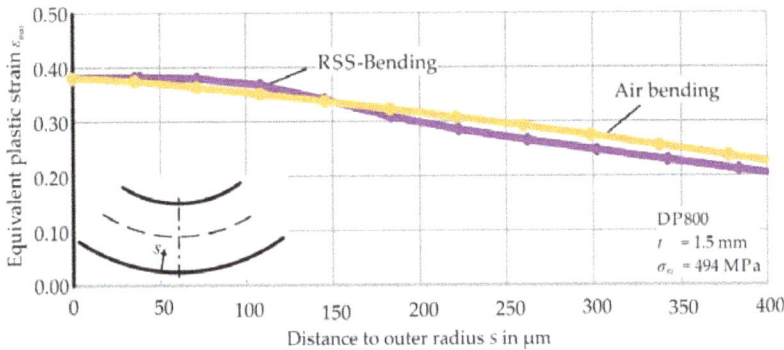

Figure 14. Equivalent strain distribution in air bending and RSS bending at the outer fiber.

In order to neglect the influence of strain hardening in the comparison of air bent and RSS-bent products, micro hardness measurements are also conducted. The average Vickers hardness HV0.1 over the bending zone at the circumference is measured as 327 ± 23 HV0.1 (η_{min} = 0.57) for air bending and 326 ± 19 HV0.1 (η_{min} = −0.06) for RSS bending. The Vickers measurements are conducted with an HMV-G21D hardness testing machine (Shimadzu Corporation, Kyoto, Japan) and a test load of 980.7 mN. The duration time of indention is 10 s and 40 indentations with a distance of 0.075 mm between the measuring were done per sample. For each bending process, two samples were investigated.

Due to this neglectable difference in sample geometry and applied strains, the undergone stress state during forming can be thought to be responsible for any measured deviations in damage characteristics. The stress triaxiality values calculated over the sheet thickness in the outer fiber differ from air bending to RSS bending (Figure 15).

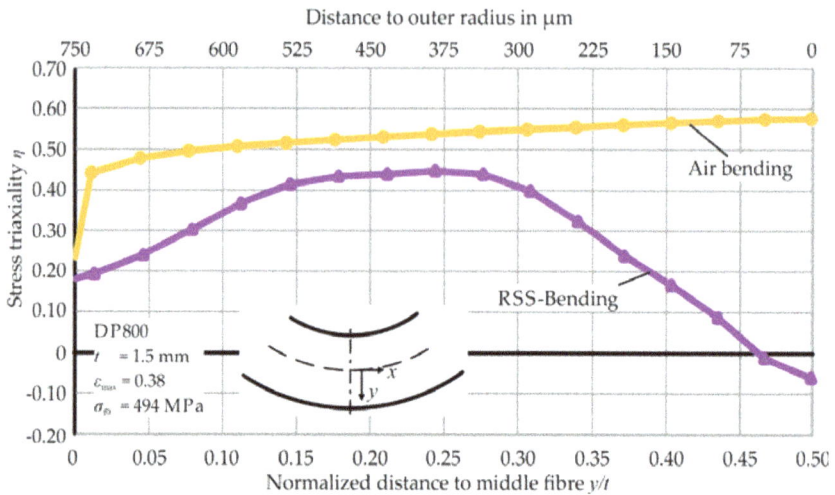

Figure 15. Triaxiality distribution in air bending and RSS bending during maximum stress superposition in the outer fibers.

In general, the air bending manifests a triaxiality of $\eta = 0.57$ at the outer fiber. In RSS bending, stress triaxialities are lower for every point at the outer fiber while superposed stresses are applied by the bending tool. During the superposition of compressive stresses, triaxiality reaches negative values at the outer fiber ($\eta_{min} = -0.06$), where the highest void volume fraction is expected due to the highest plastic strains. After the point of maximum stress superposition (Figure 15), triaxiality increases to $\eta = 0.57$ at the outer fiber until the plastic strain remains constant due to the moving forming zone in RSS bending [37].

4. Discussion

In the direct SEM-based observations of void evolution in bending samples, all classically known mechanisms and patterns of damage formation in dual-phase steels could be observed. The samples did not yet enter the state of crack formation, which is essential for observing and evaluating single voids and establishing a clear distinction between damage formation and material failure.

Deformation-induced damage voids can be sighted for regions up to 500 μm under the outer surface; for the area further towards the middle of the sheet, tensile strains are not high enough to nucleate a measurable number of voids. However, a dominant result of the void area measurements is that the average void size over all voids observed in a region does not increase drastically. This is explained by the fact that the nucleation of voids does not stop at higher plastic equivalent strains, leading to a steadily increasing number of newly nucleated, small voids. These are as dominant for the evolution of average void sizes as the evolution and growth of a few, single voids and therefore overshadow the growth of single voids when regarding mean void sizes only. Contrasting the observations of single, significantly larger voids near the outer radius, however, no increase in mean void size is observed for the RSS-bending process.

In comparison with the conventional air-bending process, the reduced triaxialities in the radial stress superposed bending process lead to a significant decrease in void area fractions. Compared to mean void size calculations, regarding area fractions yields a more complete picture of the damage state, as it takes both void nucleation and growth into account in its cumulative approach. This effect is particularly underlined by the previously-mentioned pronounced growth of voids that can be seen in the samples deformed with the conventional air-bending process. As shown in Figure 11, a clear evolution pattern of voids in a radial direction towards the outer radius could be observed; starting with mainly small voids, typically in the form of martensite cracking of void nucleation in martensite bands, the increasing tensile strains in all bending samples lead to a pronounced growth and additional nucleation of voids in that direction. Up to the outer radius of the sample, where significantly larger, evolved voids are observed, a considerate difference in void evolution behavior could be observed, as incidents of largely evolved voids over 2 μm^2 in area are solely found in the air-bending samples, and not in the RSS-bent samples. This behavior of damage evolution is explained with the reduced triaxiality depicted in Figure 15 for the RSS-bending process. Corresponding to the damage model of Oyane [38], the measurements show a decrease in void area fraction for lower triaxialities in the RSS-bending process and equally in the maximum observed void sizes of 2.26 μm^2 compared to 1.22 μm^2.

The area fractions observed for the cross-section measurements range significantly under the in-plane measurements, but still show a clear distinction between air- and RSS-bending samples. As plane strain bending can be assumed in the middle of the sheet, no macroscopic strains in any of the plane directions for this observation plane occur. Here, tensile strains are perpendicular to the observation plane. The void morphology in the in-plane measurements was dominated by the growth of voids in the x-direction, this being the direction of tensile strains. Therefore, the growth in size of these voids is not to be observed in a cross-section of the sample. Void area fractions are therefore expected to be significantly higher for in-plane measurements as the major part of the void growth can be observed and measured.

The density measurements reveal a decrease in density of 0.15% for air bending and 0.10% for RSS bending compared to the unbent material. This implies an increase of void volume fraction by 33% for the air-bending process. Compared to the results of the undergone SEM analysis, values in a comparable magnitude are calculated: for RSS bending, the average decrease in void area fraction is calculated to be 44.7%.

Compared to previous work on bending processes, a drastic improvement in both damage-reduced bending technology and damage quantification is realized. Solid counter punches [15] as well as elastomer bending [17] reduce the tendency of cracking and reduce damage; however, these effects on damage void formation and growth have so far not been able to be quantified in a detailed way. The predicted damage reduction due to compressive stress superposition according to Lemaitre [3] has been calculated for elastomer bending [39]. In this work, the applied advanced characterization methods enable the experimental determination of damage quantity and its correlation to stress triaxiality. Additionally, in contrast to the aforementioned bending processes, RSS bending does not only reduce triaxiality and therefore damage, but also proves more controllable and reproducible.

5. Conclusions

A quantitative approach for the characterization of forming-induced damage is mandatory for accurately estimating product performance.

High-resolution SEM imaging coupled to automated void recognition has proven to enable the area measurement of a statistically significant proportion of microstructural, forming-induced voids. This leads to a large-area observation of void sizes and subsequent calculation of void area fractions. A detailed analysis of all occurring voids is therefore made possible.

Void area fractions as well as density measurements show a decrease in the same order of magnitude for damage quantity in the RSS-bending process, which is correlated to its lower triaxiality values, as differences in plastic strain are negligible.

Both the overall number of forming-induced voids and their maximum size is measured to be affected by the altered stress state. The nucleation as well as the growth of voids is therefore assumed to be dependent on the magnitude of triaxiality.

Compared to other bending processes that use superposed stresses to increase formability, the RSS-bending process has proven to also reduce microstructural damage occurring before the onset of fracture without their typical loss in reproducibility or restrictions in the magnitude of superposed stresses.

Designing geometrically identical bent parts using alternative load paths has proven to reduce damage, which will consequently lead to an increased performance. This will contribute to lightweight design via the realization of thinner sheet metal parts while maintaining their mechanical performance.

Author Contributions: R.M. performed the bending experiments and analyzed the data of the bending processes and the density measurements. C.F.K. performed the SEM-measurements and developed the methodology of automatic SEM-analysis. R.M. and C.F.K. wrote the manuscript. C.L., T.A.-S., S.K.-K. and A.E.T. reviewed the results and conclusions.

Funding: The investigations are kindly supported by the German Research Foundation in context of the Collaborative Research Centre CRC/Transregio 188 "Damage-Controlled forming processes", projects A05 and B02.

Acknowledgments: The investigations are kindly supported by the German Research Foundation in context of the Collaborative Research Centre CRC/Transregio 188 "Damage-Controlled forming processes", projects A05 and B02. We thank Mr. Michael Breitwieser for the density measurements and his lecture of the density theory.

Conflicts of Interest: The authors declare no conflict of interest.

References

1. Ghadbeigi, H.; Pinna, C.; Celotto, S.; Yates, J.R. Local plastic strain evolution in a high strength dual-phase steel. *Mater. Sci. Eng. A* **2010**, *527*, 5026–5032. [CrossRef]
2. Tasan, C.C.; Diehl, M.; Yan, D.; Bechtold, M.; Roters, F.; Schemmann, L.; Zheng, C.; Peranio, N.; Ponge, D.; Koyama, M.; et al. An Overview of Dual-Phase Steels: Advances in Microstructure-Oriented Processing and Micromechanically Guided Design. *Annu. Rev. Mater. Res.* **2015**, *45*, 391–431. [CrossRef]
3. Lemaitre, J. A Continuous Damage Mechanics Model for Ductile Fracture. *J. Eng. Mater. Technol.* **1985**, *107*, 83–89. [CrossRef]
4. Gurson, A.L. Continuum Theory of Ductile Rupture by Void Nucleation and Growth: Part I—Yield Criteria and Flow Rules for Porous Ductile Media. *J. Eng. Mater. Technol.* **1977**, *99*, 2–15. [CrossRef]
5. Tasan, C.C.; Diehl, M.; Yan, D.; Zambaldi, C.; Shanthraj, P.; Roters, F.; Raabe, D. Integrated experimental—Simulation analysis of stress and strain partitioning in multiphase alloys. *Acta Mater.* **2014**, *81*, 386–400. [CrossRef]
6. McClintock, F.A. A Criterion for Ductile Fracture by the Growth of Holes. *J. Appl. Mech.* **1968**, *35*, 363–371. [CrossRef]
7. Mukherjee, K.; Hazra, S.S.; Militzer, M. Grain Refinement in Dual-Phase Steels. *Metall. Mater. Trans. A* **2009**, *40*, 2145–2159. [CrossRef]
8. Kadkhodapour, J.; Butz, A.; Ziaei Rad, S. Mechanisms of void formation during tensile testing in a commercial, dual-phase steel. *Acta Mater.* **2011**, *59*, 2575–2588. [CrossRef]
9. Archie, F.; Li, X.; Zaefferer, S. Micro-damage initiation in ferrite-martensite DP microstructures: A statistical characterization of crystallographic and chemical parameters. *Mater. Sci. Eng. A* **2017**, *701*, 302–313. [CrossRef]
10. Landron, C.; Bouaziz, O.; Maire, E.; Adrien, J. Characterization and modeling of void nucleation by interface decohesion in dual phase steels. *Scr. Mater.* **2010**, *63*, 973–976. [CrossRef]
11. Erdogan, M. The effect of new ferrite content on the tensile fracture behaviour of dual phase steels. *J. Mater. Sci.* **2002**, *37*, 3623–3630. [CrossRef]
12. Lai, Q.; Bouaziz, O.; Gouné, M.; Brassart, L.; Verdier, M.; Parry, G.; Perlade, A.; Bréchet, Y.; Pardoen, T. Damage and fracture of dual-phase steels: Influence of martensite volume fraction. *Mater. Sci. Eng. A* **2015**, *646*, 322–331. [CrossRef]
13. Bai, Y.; Wierzbicki, T. A new model of metal plasticity and fracture with pressure and Lode dependence. *Int. J. Plast.* **2008**, *24*, 1071–1096. [CrossRef]

14. Anderson, D.; Butcher, C.; Pathak, N.; Worswick, M.J. Failure parameter identification and validation for a dual-phase 780 steel sheet. *Int. J. Solids Struct.* **2017**, *124*, 89–107. [CrossRef]

15. Cupka, V.; Nakagava, T.; Tiyamoto, H. Fine bending with Counter Pressure. *Ann. CIRP* **1973**, *22*, 73–74.

16. Gänsicke, B. Verbesserung des Formänderungsvermögen bei der Blechumformung Mittels Partiell Überlagerter Druckspannung. Ph.D. Thesis, Ruhr Universität Bochum, Bochum, Germany, 2002.

17. Schiefenbusch, J. Untersuchungen zur Verbesserung des Umformverhaltens von Blechen beim Biegen. Ph.D. Thesis, Universität Dortmund, Dortmund, Germany, 1983.

18. Meya, R.; Löbbe, C.; Tekkaya, A.E. Stress State Control by a novel bending process and its effect on damage evolution. In Proceedings of the 2018 Manufacturing Science and Engineering Conference MSEC, College Station, TX, USA, 18–22 June 2018.

19. Tekkaya, A.E.; Ben Khalifa, N.; Hering, O.; Meya, R.; Myslicki, S.; Walther, F. Forming-induced damage and its effects on product properties. *CIRP Ann. Manuf. Technol.* **2017**, *66*, 281–284. [CrossRef]

20. Meya, R.; Löbbe, C.; Hering, O.; Tekkaya, A.E. New bending process with superposition of radial stresses for damage control. In Proceedings of the Forming Technology Forum, Enschede, The Netherlands, 12–13 October 2017.

21. Lemaitre, J.; Dufailly, J. Damage measurements. *Eng. Fract. Mech.* **1987**, *28*, 643–661. [CrossRef]

22. Tasan, C.C.; Hoefnagels, J.P.M.; Geers, M.G.D. Identification of the continuum damage parameter: An experimental challenge in modeling damage evolution. *Acta Mater.* **2012**, *60*, 3581–3589. [CrossRef]

23. Samuels, L.E. The nature of mechanically polished metal surfaces: The surface deformation produced by the abrasion and polishing of 70: 30 brass. *Wear* **1957**, *1*, 261. [CrossRef]

24. Zhong, Z.; Hung, N.P. Grinding of alumina/aluminum composites. *J. Mater. Process. Technol.* **2002**, *123*, 13–17. [CrossRef]

25. Isik, K.; Gerstein, G.; Clausmeyer, T.; Nürnberger, F.; Tekkaya, A.E.; Maier, H.J. Evaluation of Void Nucleation and Development during Plastic Deformation of Dual-Phase Steel DP600. *Steel Res. Int.* **2016**, *87*, 1583–1591. [CrossRef]

26. Maire, E.; Bouaziz, O.; Di Michiel, M.; Verdu, C. Initiation and growth of damage in a dual-phase steel observed by X-ray microtomography. *Acta Mater.* **2008**, *56*, 4954–4964. [CrossRef]

27. Ratcliffe, R.T. The measurement of small density changes in solids. *Br. J. Appl. Phys.* **1965**, *16*, 1193–1196. [CrossRef]

28. Schmitt, J.H.; Jalinier, J.M.; Baudelet, B. Analysis of damage and its influence on the plastic properties of copper. *J. Mater. Sci.* **1981**, *16*, 95–101. [CrossRef]

29. Bompard, D.P. Effets Endommageants de la Porosité sur la Propagation des Fessiures Dans le Nickel Fritté. Ph.D. Thesis, Universite de Technologie de Compiegne, Compiegne, France, 1986.

30. Lapovok, R. Damage evolution under severe plastic deformation. *Int. J. Fract.* **2002**, *115*, 159–172. [CrossRef]

31. Vedaldi, A.; Fulkerson, B. Vlfeat. In Proceedings of the International Conference on Multimedia—MM 2010, Firenze, Italy, 25–29 October 2010; ACM Press: New York, NY, USA, 2010; p. 1469.

32. Vincent, L.; Soille, P. Watersheds in digital spaces: An efficient algorithm based on immersion simulations. *IEEE Trans. Pattern Anal. Mach. Intell.* **1991**, *13*, 583–598. [CrossRef]

33. Kusche, C.; Reclik, T.; Freund, M.; Al-Samman, T.; Kerzel, U.; Korte-Kerzel, S. High-resolution, yet statistically relevant, analysis of damage in DP steel using artificial intelligence. *arXiv*, 2018; arXiv:1809.09657.

34. Rappoport, Z. *CRC Handbook of Tables for Organic Compund Identification*; CRC Press: Boca Raton, FL, USA, 1967.

35. Breitwieser, M. Bestimmungsmethoden der Dichte—IMETER. Available online: www.imeter.de (accessed on 15 January 2019).

36. Akeret, R. Versagensmechanismen beim Biegen von Aluminiumblechen und Grenzen der Biegefähigkeit. *Aluminium* **1978**, *54*, 117–123.

37. Meya, R.; Löbbe, C.; Tekkaya, A.E. Stress state analysis of radial stress superposed bending. *Int. J. Precis. Eng. Manuf.* **2018**, *20*, 53–66. [CrossRef]

38. Oyane, M.; Sato, T.; Okimoto, K.; Shima, S. Criteria for ductile fracture and their applications. *J. Mech. Work. Technol.* **1980**, *4*, 65–81. [CrossRef]

39. El Budamusi, M.; Becker, C.; Clausmeyer, T.; Gebhard, J.; Chen, L.; Tekkaya, A.E. *Erweiterung der Formänderungsgrenzen von Höherfesten Stahlwerkstoffen bei Biegeumformprozessen Durch Innovative Prozessführung und Werkzeuge*; IGF-Nr. 16585 N/FOSTA P930; Verl. Und Vertriebsges. Mb: Düsseldorf, Germany, 2015.

Article

A Mathematical Model of Deformation under High Pressure Torsion Extrusion

Roman Kulagin [1,*], Yan Beygelzimer [1,2], Yuri Estrin [3,4], Yulia Ivanisenko [1], Brigitte Baretzky [1] and Horst Hahn [1]

[1]   Institute of Nanotechnology, Karlsruhe Institute of Technology, 76344 Eggenstein-Leopoldshafen, Germany; yanbeygel@gmail.com (Y.B.); julia.ivanisenko@kit.edu (Y.I.); brigitte.baretzky@kit.edu (B.B.); horst.hahn@kit.edu (H.H.)

[2]   Donetsk Institute for Physics and Engineering named after A.A. Galkin, National Academy of Sciences of Ukraine, 03680 Kyiv, Ukraine

[3]   Department of Materials Science and Engineering, Monash University, Clayton 3800, Australia; yuri.estrin@monash.edu

[4]   Department of Mechanical Engineering, The University of Western Australia, Crawley 6009, Australia

[*]   Correspondence: roman.kulagin@kit.edu; Tel.: +49-721-6082-8127

Received: 16 January 2019; Accepted: 4 March 2019; Published: 8 March 2019

Abstract: High pressure torsion extrusion (HPTE) is a promising new mechanism for severe plastic deformation of metals and alloys. It enables the manufacture of long products with a radial gradient ultrafine-grained structure and of composite materials with a helical inner architecture at the meso and the macro scale. HPTE is very promising as a technique enabling light weighting, especially with magnesium, aluminium and titanium alloys. For the first time, this article presents an analytical model of the HPTE process that makes it possible to investigate the role of the various process parameters and calculate the distribution of the equivalent strain over the entire sample length. To verify the model, its predictions were compared with the numerical simulations by employing the finite element software QForm. It was shown that potential negative effects associated with the slippage of a sample relative to the container walls can be suppressed through appropriate die design and an efficient use of the friction forces.

Keywords: light metals; processing; severe plastic deformation; high pressure torsion extrusion; finite element model; equivalent strain; mechanical properties

---

1. Introduction

Recent years have seen a growth in popularity of the concept of architectured materials, which makes potential breakthroughs in materials science a realistic possibility [1–4]. This has prompted innovative developments in various areas of materials engineering and design [4]. In particular, the emerged paradigm of architectured materials offered an invigorating stimulus to the area of severe plastic deformation (SPD) technologies [5,6]. Not only can these technologies produce a submicron scale grain structure that provides metals and alloys with exceptional mechanical performance, but they can also be used to form various inner architectures of a workpiece at a mesoscopic and macroscopic scale [7–11]. In this regard, the SPD methods involving plastic torsion, such as high pressure torsion (HPT) [12,13], incremental HPT [14], twist extrusion [15], shear extrusion [16], high pressure torsion extrusion (HPTE) [17,18], torsion extrusion [19], three roll planetary milling [20], spiral equal channel angular extrusion [21], tandem process of simple shear extrusion, and twist extrusion [22]. Along with producing a submicron scale structure in the processed materials, these methods make it possible to form a helical architecture of various inclusions or reinforcements introduced in a workpiece beforehand [7–9,11,23]. These kinds of structures often occur in nature [24], which suggests that they may be very beneficial and promising in materials engineering as well. In particular, ultrafine-grained (UFG) materials with helical inner architecture offer themselves for applications of magnesium, aluminium, and titanium alloys in lightweight structures with high specific strength [9].

Multi-scale structures produced by SPD techniques are controlled by deformation processes. Therefore, for obtaining submicron structured architectured materials with desired characteristics, specific parameters of the SPD process need to be established. In some cases, the deformation of a sample can be effectively controlled through the appropriate choice of the tool geometry and/or the regime of the sample motion. The major problem is that the SPD processes mentioned above are not controllable in this way. Undesirable slippage of the sample at the contact surface with the container may cause non-steadiness of these processes, thus making them ill-controlled. For example, slippage during HPT causes a decrease in strain and its non-uniform distribution over the sample thickness [25,26]. A mathematical model [27] that accounts for slippage makes it possible to predict such effects and design the process accordingly.

A detrimental effect of slippage also occurs during HPTE. In essence, in this process a cylindrical sample is moved through two joined coaxial containers, one of which is stationary and the other is rotated about its axis [17]. The contact friction force in the second container produces rotation of the sample, which is opposed by the friction force in the first one. For sufficiently long portions of the sample in each of the two containers, these forces are large enough to produce a torque that is necessary for plastic deformation of the material. When these portions are not long enough, the sample slips in the tangential direction, the character of the process becomes non-steady and the equivalent von Mises strain at its initial and final stages is reduced.

HPTE combines the benefits of HPT with the ability to process long samples in a semi-continuous way—a quality the conventional HPT does not possess. This makes HPTE interesting for industrial scale applications. For better control of the HPTE process, a container design with special holding elements was suggested [18]. For sufficiently large dimensions, this design ensures full suppression of sample slippage. However, long holding elements increase the non-uniformity of deformation of the end parts of the sample, which is undesirable. Thus, one is faced with the problem of finding a reasonable compromise between the conflicting requirements of maximizing the friction forces and minimizing the proportion of the non-uniformly deformed portion of the sample length.

Another crucial aspect of design of the HPTE process is ensuring that simple shear deformation occurs in a thin layer of the sample [28–30]. To solve these problems, we propose a simple mathematical model that makes it possible to calculate the optimal parameters of the HPTE process. The model is presented in the subsequent sections.

2. The Model

The principal features of the HPTE process are schematically illustrated in Figure 1, which also defines the quantities used in the calculations of the process.

In Figure 1c, the plane where the two joined containers (not shown in the drawing) meet is denoted by S. A cylindrical part of height H indicates a zone in which the holding elements are located. We assume that at any time the sample consists of two stiff blocks: block 1 sitting in the stationary upper container and block 2 residing in the lower container that is rotated with an angular velocity ω. Owing to tangential slippage of the sample relative to the walls of the containers, the angular velocities ω_1 and ω_2 of the rotation of the blocks differ from those of the containers. The sample is translated downwards with a speed ϑ, so that the length of the first block is decreased and that of the second block is increased. The lengths of the parts of blocks 1 and 2 that lie outside of the holding elements are denoted L_1 and L_2, respectively. Obviously, they are related through $L_1 + L_2 = L - H$, where L is the total sample length. The kinematics of the HPTE thus defined imposes rotation of blocks 1 and 2 with angular velocities ω_1 and ω_2. These are the quantities we set out to calculate.

Friction between the sample and the container walls gives rise to a stress $\tau = mk$, where k is the shear yield stress of the material and m is the coefficient of plastic friction. The plastic friction is assumed since large pressures are required to obtain high quality submicron microstructures, in which case such a model is more consistent with experiment than the classical Coulomb friction law [31].

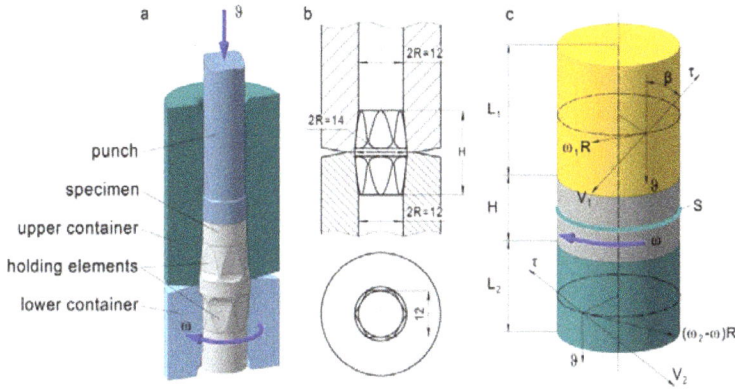

Figure 1. High pressure torsion extrusion: (**a**) schematics of the process, (**b**) design of containers and (**c**) definitions of the quantities for calculations. (unit: mm).

The material of the sample is assumed to be rigid-plastic and non-strain hardening. These assumptions are in keeping with the velocity field that experiences a discontinuity postulated below. We note that a strain hardening material does not allow for discontinuities. As was shown earlier (see, e.g., [31]), these assumptions yield meaningful estimates of the equivalent strain.

Expressions for the torques associated with the friction forces acting on blocks 1 and 2 can be obtained in the following way. According to solid mechanics, an elementary torque dM_f acting on a surface area dS of a sample is given by $dM_f = R \cdot \tau dS \cdot \sin\beta$, where β is the angle between the direction of the friction force and the rotation axis and R is the radius of the cylindrical channels of the containers. Considering that at any point on the sample surface the friction force is directed against the direction of movement of that point relative to the container surface and using $\sin\beta$ as determined from Figure 1, the friction torques M_{f1} and M_{f2} acting on blocks 1 and 2 are obtained upon integration of the elementary torques over the surfaces of the two blocks:

$$M_{f1} = 2\pi L_1 R^2 mk \frac{w_1 R}{\sqrt{\vartheta^2 + (w_1 R)^2}} \qquad (1)$$

$$M_{f2} = 2\pi L_2 R^2 mk \frac{(w - w_2)R}{\sqrt{\vartheta^2 + (w - w_2)R)^2}} \qquad (2)$$

The torques, produced by the upper and lower holding elements, M_{h1} and M_{h2}, are assumed to be equal: $M_{h1} = M_{h2} = M_h$. Their magnitude M_h is presented in the form $M_h = CM_T$, where

$$M_T = \frac{2}{3}\pi k R^3 \qquad (3)$$

is the torque associated with the plastic twisting of the sample [31]; C is a proportionality coefficient.

We further assume that for a given design of the holding elements, the magnitude of the torque they exert on the sample is proportional to their combined length, that is, that the relation $M_{h1} = (H/H_0)M_T$ holds, where H_0 is the smallest length of the holding elements with the same design, which produce the torque M_T.

The torques associated with friction forces at the end faces of the sample are neglected. Under this assumption the torque on the sample is underestimated, since in the real process the friction on its end surfaces produces torques of opposite sense in the upper and lower containers. As a result, the length of the deformed part of the sample is underestimated as well.

The magnitude of the resultant torques in the upper and lower containers is given by $M_{f1} + M_h$ and $M_{f2} + M_h$ respectively.

Let us consider two possible cases: (i) $w_1 = w_2 = w_0$ and (ii) $w_2 > w_1$.

In the first case, the sample moves as a whole, meaning that no plastic deformation occurs. In the second case, the tangential component of the velocity experiences a discontinuity at the plane where the two containers meet. Plasticity theory [31] tells us that this leads to simple shear in the material corresponding to a von Mises strain $e_M = \frac{[\vartheta_\tau]}{\sqrt{3}\vartheta_n}$. Here $[\vartheta_\tau]$ is the absolute value of the jump in the tangential component of the velocity and is the absolute value of the velocity component normal to the plane of discontinuity. In our case, the relations $[\vartheta_\tau] = r(w_2 - w_1)$ and $\vartheta_n = \vartheta$ hold. Substitution of these expression in the above formula for the von Mises equivalent strain yields

$$e_M = \frac{1}{\sqrt{3}} \cdot \frac{r(w_2 - w_1)}{\vartheta} \tag{4}$$

Variant (i) is realized when the torque applied to the sample is smaller than the torque associated with plastic torsion. In that case only the equality of the absolute values of the torques acting on the sample in the two containers, $M_{f1} + M_h = M_{f2} + M_h$, must hold, which can be re-written as

$$L_1 \frac{w_1}{\sqrt{\vartheta^2 + (w_1 R)^2}} = L_2 \frac{(w - w_2)}{\sqrt{\vartheta^2 + ((w - w_2)R)^2}} \tag{5}$$

We now use the equality $w_1 = w_2 = w_0$ in Equation (5) and obtain the following implicit relation for the normalized velocity $a = \vartheta/(wR)$ of the rotation of the entire sample:

$$\left(\frac{1}{\xi} - 1\right) \frac{w_0'}{\sqrt{a^2 + w_0'^2}} = \frac{1 - w_0'}{\sqrt{a^2 + (1 - w_0')^2}} \tag{6}$$

where $\xi = L_2/(L - H)$ and $w_0' = w_0/w$.

Variant (ii) is realized if the condition

$$\begin{cases} M_{f1} + M_h = M_T \\ M_{f2} + M_h = M_T \end{cases} \tag{7}$$

is fulfilled. In this case the angular velocities of blocks 1 and 2 can be determined from a set of two equations: the torque equilibrium equation, Equation (6), and the relation expressed by Equation (7), which, after substitution of the expressions for the torques takes the form

$$2\pi L_2 R^2 mk \frac{(w - w^2)R}{\sqrt{\vartheta^2 + ((w - w^2)R)^2}} + M_h = \frac{2}{3}\pi k R^3 \tag{8}$$

Equation (8) can be re-written as

$$\xi m \frac{(1 - w_2')}{\sqrt{a^2 + (1 - w_2')^2}} = \frac{1}{3}\frac{R}{(L - H)} - \frac{M_h}{2\pi k R^2(L - H)} = \frac{1}{3}\frac{R}{(L - H)}(1 - H') \tag{9}$$

where $w_2' = w_2/w$ and $H' = H/H_0$. The normalized frequency w_2' can now be found from Equation (9). By combining Equations (4) and (9) the following equation for $w_1' = w_1/w$:

$$m(1 - \xi) \frac{w_1'}{\sqrt{a^2 + w_2'^2}} = \frac{1}{3} \frac{R}{(L - H)} (1 - H') \tag{10}$$

We now wish to determine the conditions for variant (ii) to apply. To that end, we analyze Equations (9) and (10).

A non-negative solution of Equation (10) reads

$$w_1' = \frac{a \cdot b}{\sqrt{(1 - \xi)^2 - b^2}} \tag{11}$$

where the parameter b is defined by

$$b = \frac{R}{3m(L - H)} (1 - H') \tag{12}$$

It should be noted that Equation (9) is reduced to Equation (10) if the following substitutions

$$w_1' \to (1 - w_2') \text{ and } (1 - \xi) \to \xi \tag{13}$$

are made. As the condition $(1 - w_2') \geq 0$ is fulfilled, the solution of Equation (9) reads

$$w_2' = 1 - \frac{a \cdot b}{\sqrt{\xi^2 - b^2}} \tag{14}$$

From Equations (11) and (14) it follows that a solution exists only in the range of

$$b < \xi < 1 - b \tag{15}$$

As $w_2' > 0$ holds, Equation (14) puts a further restriction on the parameter ξ:

$$\frac{a \cdot b}{\sqrt{\xi^2 - b^2}} < 1 \tag{16}$$

This is tantamount to the inequality

$$\xi > b\sqrt{a^2 + 1} \tag{17}$$

Combining inequalities (15) and (17) we have

$$b\sqrt{a^2 + 1} < \xi < 1 - b \tag{18}$$

It can thus be concluded that for plastic deformation of a sample to occur, the inequality $b\sqrt{a^2 + 1} < 1 - b$ must be satisfied, i.e., the inequality

$$b < \frac{1}{1 + \sqrt{a^2 + 1}} \tag{19}$$

has to hold. Finally, substitution of the expressions defining the quantities a and b transforms inequality (19) to

$$\frac{L-H}{R} > \frac{1}{3m}(1-H')\left(1+\sqrt{\left(\frac{\vartheta}{\omega R}\right)^2+1}\right) \tag{20}$$

This inequality, expressed in terms of the container geometry and process parameters, is a basis for determining the conditions for plastic twisting of a sample under HPTE to occur.

3. Results and Discussion

We now analyse the model proposed for conditions close to real experiment [18]. The holding element design employed in [18] did ensure the occurrence of plastic torsion of samples. Let us set H_0 to be equal to the length of a holding element used in experiment ($H_0 = 24$ mm). The characteristic process parameters used in [18] and summarised in Table 1 will be adopted in the analysis to follow.

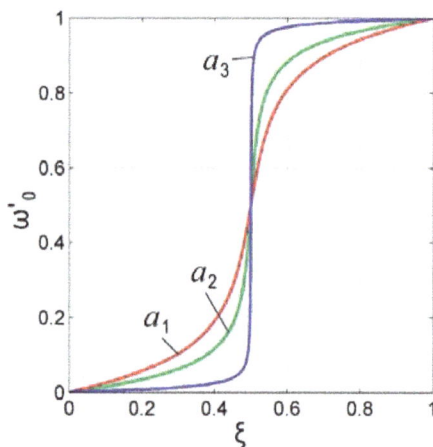

Figure 2. Dependence of the non-dimensional angular velocity of the sample, ω_0', on the position of the sample, ξ, in the container channel.

Table 1. High pressure torsion extrusion (HPTE) parameters for analysis of the model used to calculate the graphs presented in Figure 2.

Regime	ϑ, mm/min	ω, rpm	a
a_1	10	1	0.265
a_2	5	1	0.133
a_3	1	1	0.027

Figure 2 presents the dependence of ω_0' on ξ obtained by solving Equation (6) for three different values of the parameter a.

As seen from the graphs, the angular velocity of a rigid sample translated through the joint plane between the containers varies from zero (when the entire sample sits in the stationary upper container) to ω—the angular velocity of rotation of the lower container. The smaller the magnitude of the parameter a, the more precipitous is this transition from zero to ω at the point in time when the middle of the sample passes the joint plane between the two containers ($\xi = 0.5$).

The dependence of the angular velocities obtained by solving Equations (9) and (10) on the process parameters is presented graphically in Figure 3, which depicts ω_1' and ω_2' vs. ξ.

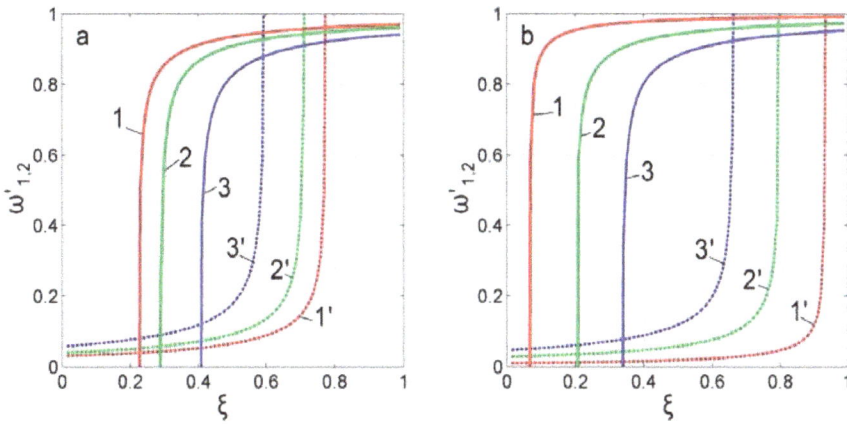

Figure 3. Dependence of the normalized angular velocities of the two parts of a sample (blocks 1 and 2), ω'_1 (1, 2, 3) and ω'_2 (1′, 2′, 3′) on the position of the sample in the HPTE container channel, ξ, for (**a**) different values of the coefficient of friction ($m = 0.9$ (1 and 1′); $m = 0.7$ (2 and 2′); $m = 0.5$ (3 and 3′)) in the absence of holding elements and (**b**) different lengths of the holding elements ($H' = 0.9$ (1 and 1′); $H' = 0.7$ (2 and 2′); $H' = 0.5$ (3 and 3′)) for a fixed magnitude of the coefficient of friction ($m = 0.3$).

As discussed above, plastic deformation of the sample only occurs for the range of parameters where the double inequality (18) is fulfilled. This corresponds to the region between the two intercepts of the curves for ω'_1 and ω'_2 in Figure 3. The first intercept determines the beginning of the plastic twisting of the sample. As the sample moves through the container, the angular velocities of block 1 and block 2 rise. Initially, ω'_2 grows faster, but, starting from $\xi = 0.5$, the growth of ω'_1 becomes faster than that of ω'_2. Eventually, at the second intercept of the curves ω'_1 vs. ξ and ω'_2 vs. ξ, plastic twisting comes to a standstill.

Figure 3a illustrates the capacity of frictional forces with regard to the feasibility of HPTE without holding elements. For a fixed coefficient of friction ($m = 0.3$) the holding elements enable a longer part of a sample to undergo deformation by simple shear, as evident from a comparison between Figure 3a,b.

To verify the quality of the proposed mathematical model, we compare the distribution of the equivalent strain over the sample length following from Equation (5) with that calculated using the finite element software package QForm [32]. Two variants of the computations will be considered. To do this, consider two options for calculating. The first one corresponds to a real experiment conducted on copper samples [18]: $L = 35$ mm, $H = 24$ mm, $H_0 = 24$ mm, $R = 6$ mm, $m = 0.15$, $\omega = 1$ rpm and $\vartheta = 10$ mm/min. In the second variant, shorter holding elements ($H = 12$ mm) and a larger coefficient of friction ($m = 0.6$) are employed in the calculations. Both variants were simulated using finite element method (FEM) modelling. The upper and the bottom containers as well as the punch were defined as rigid bodies, whereas the workpiece was represented by 30,000 tetragonal deformable elements. Adaptive meshing was also used for the workpiece. The material behavior was taken to be isotropic using the von Mises perfect plasticity model. Friction between the billet and the container walls was modelled by the plastic friction law.

Perfect plasticity was used both in the analytical model and in the FEM simulations. This was done to provide a common ground for both modelling approaches and enable a comparison between the predictions of the two models. Furthermore, in [23] the authors looked into the influence of the choice of the material model on the deformation state of the workpiece and showed that the details of the constitutive law can be neglected in a first approximation. In this connection it should also be mentioned that experimental data for SPD exhibit a trend of the mechanical properties towards a steady state, thus justifying perfect plasticity as an adequate approach. One should recognise, however, that in some cases (see, e.g., [33]) the rheology can have a strong influence on the deformation state of the workpiece leading to such effects as strain localisation. Investigations of such effects are enabled by simple rheological models, as suggested, for example, in [34,35]. An analysis of the role of rheology in the deformation behaviour of a workpiece under HPTE processing will be a subject of future research.

The equivalent strain distributions in the transversal cross-section through the middle of the sample for five consecutive points in time are presented in Figure 4.

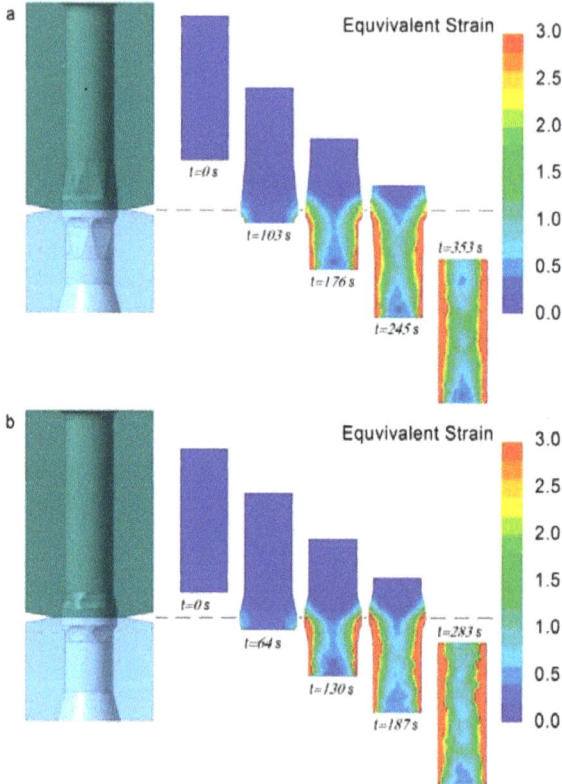

Figure 4. The equivalent strain distributions in the transversal cross-section through the middle of the sample calculated by (**a**) variant 1 and (**b**) variant 2 by means of QForm.

Figure 4 illustrates the accumulation of strain progressing with the movement of the sample through the holding elements. It clearly shows the non-uniformity of strain across the bulk of the sample. The dependence of the non-uniformity on the sample radius is associated with the increase of strain with the distance from the axis to the periphery of the sample, which is inherent in the torsion-based processes and is common to HPT and HPTE. Furthermore, near the surface the sample undergoes additional deformation through the cusps on the holding elements.

Non-uniformity of strain over the length of a sample is associated with a gradual increase of the counter-torque with the sample continually filling the holding elements, compounded with an insufficient capacity of the frictional forces at the beginning and the end of the process. The two variants considered differ in the contribution of these factors to the formation of the longitudinal non-uniformity. In the first variant (Figure 4a), the holding elements provide the torque necessary for plastic twisting. This is enabled by their sufficiently large length, which leads to the relative size of the end zone being small. In the second variant (Figure 4b), the gripping of the sample is established by a concerted action of the frictional forces and the holding elements. As their length is relatively small, the strain non-uniformity along the sample caused by the holding element size is reduced. Given the sufficiently large capacity of the frictional forces, the uniform part of the sample is large in relation to its length.

The above statements are supported by the diagrams in Figure 5 which display equivalent strain distributions over the length of a sample at $r = 0.5R$.

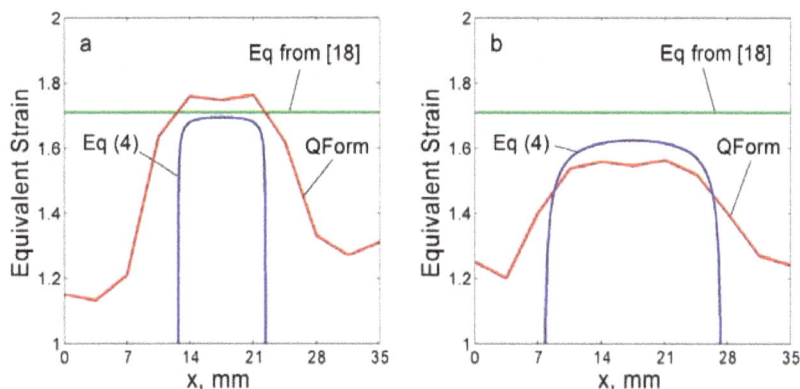

Figure 5. Dependence of the equivalent strain on the distance x from the sample end surface at $r = 0.5R$: (**a**) variant 1 and (**b**) variant 2. The calculations done by using Equation (4) from the present work and the finite element simulations with QForm are seen to compare reasonably well. A constant level of equivalent strain calculated in [18] is also shown as a reference.

As seen in Figure 5, utilizing the friction force capacity in combination with the use of holding elements enables broadening of the uniformly deformed part of the sample. However, the magnitude of the equivalent strain in this portion of the sample is somewhat reduced. This is caused by slippage of the sample, which occurs in the case presented in Figure 5b. In the absence of slippage, the maximum values of the equivalent strain as predicted by the analytical model coincide with those obtained from the simple relation proposed in Reference [18] (cf. Figure 5a). The magnitude of the maximum equivalent strain in steady state computed using QForm (Micas Simulations Limited, Oxford, UK) (also shown in Figure 5) is somewhat larger. This is associated with some extra deformation in the sample parts contained within the holding elements.

Figure 5 also shows that the results obtained with the model outlined are in good agreement with the final element calculations, specifically those employing the QForm software. A precipitous drop of strain calculated using Equation (4) is caused by the model assumption that a holding element is to be fully filled with the deforming metal. In QForm, calculations are possible for partially filled holding elements, as well, which gives rise to descending parts of the graph.

Slippage of a sample relative to the containers leads to the emergence of a tangential near-surface flow of the metal within the holding elements. This gives rise to additional deformation with a shear plane aligned with the extrusion axis—an effect similar to the "cross flow" that occurs under Twist Extrusion [23]. A kinematically possible velocity field suggested in this study leads to a linear variation of strain along a sample radius. As shown in Reference [18], the strain distribution may deviate from a linear one. The current analytical modelling showed a good qualitative agreement with the results of FEM simulations with regard to the radial strain distribution, however. In addition, a good quantitative agreement was found for the regions at the periphery of a workpiece.

We consider the FEM simulation and the analytical model not as competing, but rather as being complementary to each other. The closed form analytical solution enables a large number of calculations for various combinations of the defining parameters at a low computational cost. This makes it possible to carry out initial screening as a first step in the process optimisation. This should be followed by high-precision numerical calculations as a second step of the optimisation procedure. Thus, both modelling approaches would be brought to bear in this two-step optimisation process.

4. Conclusions

In this article, a simple mathematical model of the deformation behaviour of a rigid-plastic material in a severe plastic deformation process of high pressure torsion extrusion has been presented. The model provides a prediction of the worked part of a sample, even in cases when its slippage relative to the container walls occurs. It also predicts the magnitude of the equivalent strain of the material under such conditions. It is thus believed that the mathematical relations defining the window in the parameter space where an optimum in the HPTE performance is achievable will be useful for process design. The model proposed has also demonstrated that the use of holding elements and the resource of frictional forces to create a torque on a sample that makes it possible to maximise the length of its portion where nearly steady-state working of the material is achieved. We believe that due to its simplicity, generality and robustness, the mathematical model proposed can be successfully applied to calculating the mechanical behaviour of lightweight structures made from magnesium, aluminium, and titanium.

Author Contributions: R.K. and Y.B. conceived the idea presented in the paper. Y.B. developed the theory and R.K. performed the computations. Y.E. verified the analytical model. Y.I., B.B. and H.H. supervised the project. All authors discussed the results and contributed to the final manuscript.

Funding: This research was funded by German Research Foundation (DFG) grant number IV98/8-1 and by the Federal Ministry for Economic Affairs and Energy, based on the decision of a German Bundestag, IGF grant number 19838N.

Acknowledgments: The authors acknowledge funding support from the German Research Foundation (DFG) through Grant IV98/8-1 and partially by the Federal Ministry for Economic Affairs and Energy, based on a decision of the German Bundestag, IGF grant number 19838N.

Conflicts of Interest: The authors declare no conflict of interest.

References

1. Ashby, M.F.; Brechet, Y. Designing hybrid materials. *Acta Mater.* **2003**, *51*, 5801–5821. [CrossRef]
2. Ashby, M.F. Designing architectured materials. *Scr. Mater.* **2013**, *68*, 4–7. [CrossRef]
3. Brechet, Y.; Embury, J.D. Architectured materials: Expanding materials space. *Scr. Mater.* **2013**, *68*, 1–3. [CrossRef]
4. Estrin, Y.; Brechet, Y.; Dunlop, J.; Fratzl, P. *Architectured Materials in Engineering and Nature*; Springer: Berlin, Germany, 2019; in press.
5. Valiev, R.Z.; Estrin, Y.; Horita, Z.; Langdon, T.G.; Zehetbauer, M.; Zhu, Y.T. Producing Bulk Ultrafine-Grained Materials by Severe Plastic Deformation: Ten Years Later. *JOM* **2016**, *68*, 1216–1226. [CrossRef]
6. Estrin, Y.; Vinogradov, A. Extreme grain refinement by severe plastic deformation: A wealth of challenging science. *Acta Mater.* **2013**, *61*, 782–817. [CrossRef]
7. Khoddam, S.; Estrin, Y.; Kim, H.S.; Bouaziz, O. Torsional and compressive behaviours of a hybrid material: Spiral fibre reinforced metal matrix composite. *Mater. Des.* **2015**, *85*, 404–411. [CrossRef]
8. Latypov, M.; Beygelzimer, Y.; Kulagin, R.; Varyukhin, V.; Kim, H.S. Toward architecturing of metal composites by twist extrusion. *Mater. Res. Lett.* **2015**, *3*, 161–168. [CrossRef]
9. Beygelzimer, Y.; Estrin, Y.; Kulagin, R. Synthesis of Hybrid Materials by Severe Plastic Deformation: A New Paradigm of SPD Processing. *Adv. Eng. Mater.* **2015**, *17*, 1853–1861. [CrossRef]
10. Kang, J.Y.; Kim, J.G.; Park, H.W.; Kim, H.S. Multiscale architectured materials with composition and grain size gradients manufactured using high-pressure torsion. *Sci. Rep.* **2016**, *6*, 26590. [CrossRef]
11. Beygelzimer, Y.; Kulagin, R.; Estrin, Y. Severe Plastic Deformation as a Way to Produce Architectured Materials. In *Architectured Materials in Engineering and Nature*; Estrin, Y., Brechet, Y., Dunlop, J., Fratzl, P., Eds.; Springer: Berlin, Germany, 2019; in press.

12. Bridgman, P.W. Effects of High Shearing Stress Combined with High Hydrostatic Pressure. *Phys. Rev.* **1935**, *48*, 825–847. [CrossRef]

13. Zhilyaev, A.P.; Langdon, T.G. Using high-pressure torsion for metal processing: Fundamentals and applications. *Prog. Mater. Sci.* **2008**, *53*, 893–979. [CrossRef]

14. Hohenwarter, A. Incremental high pressure torsion as a novel severe plastic deformation process: Processing features and application to copper. *Mater. Sci. Eng. A* **2015**, *626*, 80–85. [CrossRef] [PubMed]

15. Beygelzimer, Y.; Kulagin, R.; Estrin, Y.; Toth, L.S.; Kim, H.S.; Latypov, M. Twist Extrusion as a Potent Tool for Obtaining Advanced Engineering Materials: A Review. *Adv. Eng. Mater.* **2017**, *19*, 1600873. [CrossRef]

16. Segal, V. Shear-Extrusion Method. U.S. Patent 7096705 B2, 29 August 2006.

17. Fedorov, V.; Ivanisenko, J.; Baretzky, B.; Hahn, H. Vorrichtung und Verfahren zur Umformung von Bauteilen aus Metallwerkstoffen. DE Patent DE102013213072A1, 8 January 2015.

18. Ivanisenko, Yu.; Kulagin, R.; Fedorov, V.; Mazilkin, A.; Scherer, T.; Baretzky, B.; Hahn, H. High Pressure Torsion Extrusion as a new severe plastic deformation process. *Mater. Sci. Eng. A* **2016**, *664*, 247–256. [CrossRef]

19. Mizunuma, S. Large Straining Behavior and Microstructure Refinement of Several Metals by Torsion Extrusion Process. *Mater. Sci. Forum* **2006**, *503-504*, 185–190. [CrossRef]

20. Wang, Y.L.; Molotnikov, A.; Diez, M.; Lapovok, R.; Kim, H.E.; Wang, J.T.; Estrin, Y. Gradient structure produced by three roll planetary milling: Numerical simulation and microstructural observations. *Mat. Sci. Eng. A* **2015**, *639*, 165–172. [CrossRef]

21. Fadaei, A.; Farahafshan, F.; Sepahi-Boroujeni, S. Spiral equal channel angular extrusion (Sp-ECAE) as a modified ECAE process. *Mater. Des.* **2017**, *113*, 361–368. [CrossRef]

22. Kim, J.G.; Latypov, M.; Pardis, N.; Beygelzimer, Y.; Kim, H.S. Finite element analysis of the plastic deformation in tandem process of simple shear extrusion and twist extrusion. *Mater. Des.* **2015**, *83*, 858–865. [CrossRef]

23. Kulagin, R.; Latypov, M.; Kim, H.S.; Varyukhin, V.; Beygelzimer, Y. Cross Flow During Twist Extrusion: Theory, Experiment, and Application. *Metall. Mater. Trans. A* **2013**, *44a*, 3211–3220. [CrossRef]

24. Li, L.; Weaver, J.C.; Ortiz, C. Hierarchical structural design for fracture resistance in the shell of the pteropod Clio pyramidata. *Nat. Commun.* **2015**, *6*, 1. [CrossRef]

25. Edalati, K.; Horitaa, Z.; Langdon, T.G. The significance of slippage in processing by high-pressure torsion. *Scripta Mater.* **2009**, *60*, 9–12. [CrossRef]

26. Yogo, Y.; Sawamura, M.; Iwata, N.; Yukawa, N. Stress-strain curve measurements of aluminum alloy and carbon steel by unconstrained-type high-pressure torsion testing. *Mater. Des.* **2017**, *122*, 226–235. [CrossRef]

27. Khoddam, S.; Hodgson, P.D.; Zarei-Hanzaki, A.; Foon, L.Y. A simple model for material's strengthening under high pressure torsion. *Mater. Des.* **2016**, *99*, 335–340. [CrossRef]

28. Segal, V. Severe plastic deformation: Simple shear versus pure shear. *Mater. Sci. Eng. A* **2002**, *338*, 331–344. [CrossRef]

29. Segal, V. Review: Modes and Processes of Severe Plastic Deformation. *Materials* **2018**, *11*, 1175. [CrossRef] [PubMed]

30. Beygelzimer, Y.; Lavrinenko, N. Perfect plasticity of metals under simple shear as the result of percolation transition on grain boundaries. *arXiv*, 2012; arXiv:1206.5055v1.

31. Johnson, W.; Mellor, P.B. *Engineering Plasticity*; Ellis Horwood Limited; Van Nostrand Reinhold (UK) Ltd.: London, UK, 1983.

32. Metal Forming Simulation Software Qform. Available online: http://www.qform3d.com/ (accessed on 7 March 2019).

33. Beygelzimer, Y.; Kulagin, R.; Toth, L.S.; Ivanisenko, Y. The self-similarity theory of high pressure torsion. *Beilstein J Nanotechnol.* **2016**, *7*, 1267–1277. [CrossRef] [PubMed]

34. Panteghini, A. An analytical solution for the estimation of the drawing force in three dimensional plate drawing processes. *Int. J Mech. Sci.* **2014**, *84C*, 147–157. [CrossRef]

35. Panteghini, A.; Genna, F. An engineering analytical approach to the design of cold wire drawing processes for strain-hardening materials. *Int. J Mater. Form.* **2010**, *3*, 279–289. [CrossRef]

metals

MDPI

Article

Experimental and Numerical Study of an Automotive Component Produced with Innovative Ceramic Core in High Pressure Die Casting (HPDC)

Giovanna Cornacchia [1,*]**, Daniele Dioni** [1]**, Michela Faccoli** [1][iD]**, Claudio Gislon** [2]**, Luigi Solazzi** [1]**, Andrea Panvini** [1] **and Silvia Cecchel** [1][iD]

[1]   DIMI, Department of Industrial and Mechanical Engineering, University of Brescia, via Branze 38, 25123 Brescia, Italy; daniele.dioni@gmail.com (D.D.); michela.faccoli@unibs.it (M.F.); luigi.solazzi@unibs.it (L.S.); a.panvini@piq2.com (A.P.); s.cecchel@unibs.it (S.C.)

[2]   Co.Stamp. s.r.l. Via Verdi 6, 23844 Sirone (LC), Italy; claudio.gislon@costampgroup.it

*   Correspondence: giovanna.cornacchia@unibs.it; Tel.: +39-030-371-5827; Fax: +39-030-370-2448

Received: 14 December 2018; Accepted: 8 February 2019; Published: 12 February 2019

Abstract: Weight reduction and material substitution are increasing trends in the automotive industry. High pressure die casting (HPDC) is the conventional casting technology for the high volume production of light alloys; it has recently found wide application in the manufacturing of critical components, such as complex and thin geometry automotive parts. However, the major restriction of this affordable technology is the difficulty to design and realize hollow sections or components with undercuts. An innovative way to further increase the competitiveness of HPDC is to form complex undercut shaped parts through the use of new lost cores that are able endure the high pressures used in HPDC. This paper investigates the use of innovative ceramic lost cores in the production of a passenger car aluminum crossbeam by HPDC. Firstly, process and structural simulations were performed to improve the crossbeam design and check the technology features. The results led to the selection of the process parameters and the production of some prototypes that were finally characterized. These analyses demonstrate the feasibility of the production of hollow components by HPDC using ceramic cores.

Keywords: non-ferrous alloys; ceramic core; FEA; HPDC; de-coring; material characterization

---

1. Introduction

The current use of castings for aluminum chassis structural applications is limited to a niche market for high performance cars. Nevertheless, automobile weight reduction is currently becoming an essential requirement to improve vehicle performance [1] and reduce fuel consumption and harmful emissions [2–7]. The environmental importance of reducing the weight of these aluminum components has been recently demonstrated by means of a life cycle assessment tool [8–10]. The studies considered all the component life-related phases (mineral extraction, component manufacturing, use on a vehicle, and end of life), demonstrating the relevant contribution of their reduced weight to the reduction of pollutant emission during vehicle circulation. Therefore, the application of lightweight metal castings for the production of low range and high volume automotive components is increasing, not only for chassis, but also engine blocks, cylinder heads, intake manifolds, brackets, housings, transmission parts, and suspension systems [2,7–14].

Of course, the preservation of compliance with the safety and performance levels required by automotive standards is paramount [5,15]. This is particularly true for safety-relevant automotive products (i.e., crossbeams, control arms, etc.), which are defined as "reliable in regard to safety-relevant defects". Indeed, the necessity of excellent cast integrity requires the use of high cost and/or low productivity technologies, such as gravity casting or high vacuum die casting [2,16–19]. From a mechanical and structural point of view, gravity casting technology guarantees excellent performances, but it is not suitable for large volumes because the slow production cycle and component feasibility still limit the possibility of reducing the thickness to less than 3.5 mm [20–26].

In this context, a consolidated and promising technology is certainly high pressure die casting (HPDC). HPDC is a very competitive technology for high production volumes, low costs, near-net shape parts, and the opportunity to produce thin components. The main limit of HPDC technology is related to part design limitations associated with the difficulty of using lost cores, which are the only way to form complex undercut shaped parts. Indeed, the lost cores made of traditional gravity technology materials can only be utilized when the process pressure is a few bars because they are not compatible with the high hydrostatic pressure and also when the flow speed is much lower than the high flow speed involved in HPDC process. Traditionally, metallic moving cores are used when HPDC parts have undercuts. The metallic moving core, in contrast with the lost core, is a permanent mold technology [22–24]. It means that, currently, the design of HPDC hollow section has to take into account the ejection of the metallic moving core [25–30]. This limits the feasibility of automotive components with complex closed-profile sections, which offer higher torsional stiffness and a further weight reduction [5,6] and, consequently, have become more and more required in the transport field.

For all these reasons, the desire to extend HPDC to automotive components has encouraged numerous studies aimed to obtain HPDC-resistant cores. The main requirements of these lost cores are related to resistance at standard HPDC conditions, and can be broken down into mechanical properties and process reliability. As regards their mechanical properties, the principal characteristics are high Young's modulus and adequate bending and compression strengths. As regards process reliability, it is indispensable for achieving shape stability, dimensional accuracy, surface quality, thermal shock resistance, possibility of complex shapes and easy, clean removability. Different types of soluble cores have been developed in recent years [31–37].

Numerous researches have proposed many kinds of new collapsible or expendable cores comprised of sand, salt, and metallic and organic material [38–42].

In particular, the use of salt cores, already known in gravity or low pressure casting processes, has received extended attention for HPDC [39–42]. Yaokawa et al. [39] investigated the strength of four binary systems, sodium chloride–sodium carbonate, potassium chloride–potassium carbonate, potassium chloride–sodium chloride, and potassium carbonate–sodium carbonate, whose liquidus temperature is fit for use in the HPDC processing of aluminum alloys. Interesting studies on the use of lost cores made from salt (sodium chloride) were performed by Fuchs et al. [41,42]. The authors compared numerical simulation results with corresponding experiments to predict core failure during the casting process. As a result, they obtained process parameters for the successful use of this salt cores in HPDC.

Various patents have been applied for in this regard [43–48], such as for the production of aluminum closed deck cylinder blocks by HPDC [43] or aluminum based engine blocks [44]. Brown et al. [45] disclosed novel water-soluble cores for metal casting use and methods for making such cores. The cores essentially consist of a water-soluble salt and a synthetic resin. Although the literature confirms the above-reported potential for HPDC salt cores, their application for demanding automotive conditions needs additional research to reach the high strength required.

These problems were recently addressed by studing a new type of core, more precisely, an HPDC ceramic core. The choice of ceramic material is related to the use of injection molding, necessary for complex core shapes production. This technology provides good dimensional tolerance and low roughness for the internal component cavity surface. In addition, ceramic material without binder does not release gas during the casting and would consequently guarantee higher mechanical properties [49,50].

Regarding the decoring, although both leaching and high pressure water jet techniques can be easily applied for ceramic core removing, it is necessary to highlight the possible difficulty of accessing complex cavities, coupled with the particular resistance of this kind of core. This ceramic core would resist to over 1000 bar, compatible with HPDC process, instead of the maximum 2 bar of the traditional lost cores. The most relevant benefits expected from this application are reduction of machining processes, decreased production times, increased torsional stiffness, and reduction of weight.

It is against this background that the aim of this project should be seen: the study, development, and industrialisation of a new method, which uses this new ceramic core applied to HPDC. This implementation not only allows the production of a lighter and improved safety-relevant automotive component, but permits achievement of relevant benefits in comparison with both conventional die with moving cores and components made of bulk closed sections.

In this project, the feasibility study on the production of crossbeams using this innovative lost ceramic core in HPDC has demonstrated that this new type of core may contribute to making HPDC competitive with respect to the conventional casting process for the production of automotive hollow parts. The specific goal of this paper is to re-design, produce, and demonstrate the feasibility of a new, one-piece hollow aluminum crossbeam for passenger cars produced with HPDC and ceramic cores.

This paper firstly shows the results of process and structural simulations used for the selection of the most proper feature of the crossbeam. Computer aided engineering (CAE) simulation of the HPDC process was used to forecast and improve the quality of the casting and verify foundry feasibility, and finite element method (FEM) analysis was used to evaluate part performances and improve the design. Foundry process simulation has been a well-known and widely used tool for more than twenty years for improving die and process design, in order to achieve better part quality.

These investigations, supported by the experimental characterization of the ceramic cores, allowed the definition of the component geometry and the selection of the most proper materials and process parameters for the production of some crossbeam prototypes. Finally, hardness measurements and microstructural analyses were carried out on these prototypes.

2. Materials and Methods

2.1. Description of the Automotive Component Structure and Materials

2.1.1. Geometric Description

The front crossbeam is a functional, structural, and safety vehicle component that acts as a link between the suspension elements, the steering knuckles, and the main frame. Due to the high mechanical resistance required, the crossbeam of a medium-size car is traditionally produced in iron castings or trimmed and folded steel sheets.

In some recent applications [7], these heavy metals solutions were substituted by re-designed aluminum alloys parts produced with different casting technologies (i.e., HPDC) in order to reduce the weight. As stated in the Introduction section, the shape of the casting must be designed taking into account the technological limitations of HPDC, among which the most restrictive are the casting ejection from a rigid steel die and the limited number of small undercuts, which can only be managed with sliding cores.

Starting from these considerations, the new expendable ceramic core technology was applied to optimize the shape of an aluminum HPDC crossbeam in order to further reduce weight and increase stiffness. The design of either the traditional HPDC crossbeam or the modified one can be observed in Figure 1. It is worthwhile to note that the main dimensions and the material (aluminum alloy EN AC-43500) of the modified crossbeam and the original crossbeam are the same. In the new design, the central area of the component was closed to form a box section, whereas the extremities were not modified because they constitute the main connection areas with the other suspension elements and their interfaces could not be moved.

The original open profile design, sketched in Figure 1a, was modified as shown in Figure 1b, where the reinforcement ribs on the external surface, no longer necessary, were removed. The final shape of the hollow cavity is shown in Figure 2a through the expendable ceramic core that creates the cavity. In addition, in Figure 2b, it is possible to observe the geometry and the positioning solutions in the mold cavity adopted for the ceramic core. The introduction of the new ceramic core technology and the applied design modifications led to a final weight of 3.93 kg for the boxed part as compared to 3.95 kg for the traditional HPDC one.

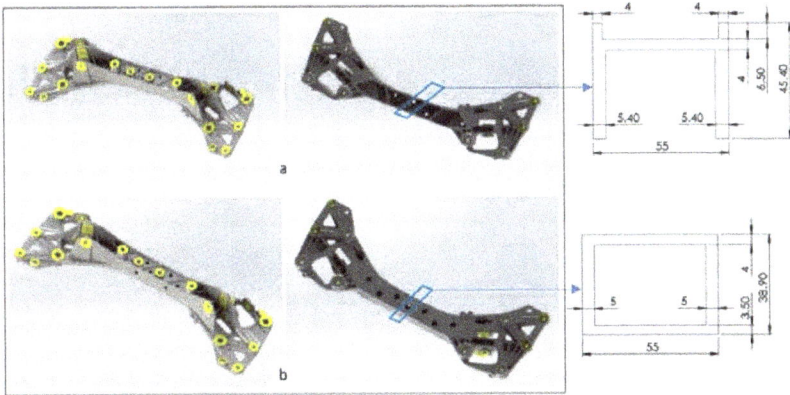

Figure 1. 3D design and schematic sketch of the section profile of (**a**) original crossbeam (**b**) modified crossbeam. The dimensions are expressed in (mm).

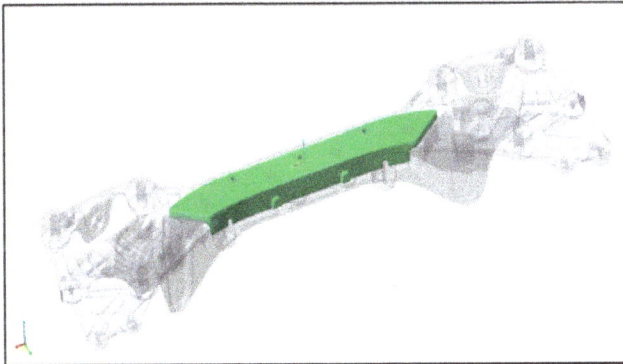

(**a**)

Figure 2. *Cont.*

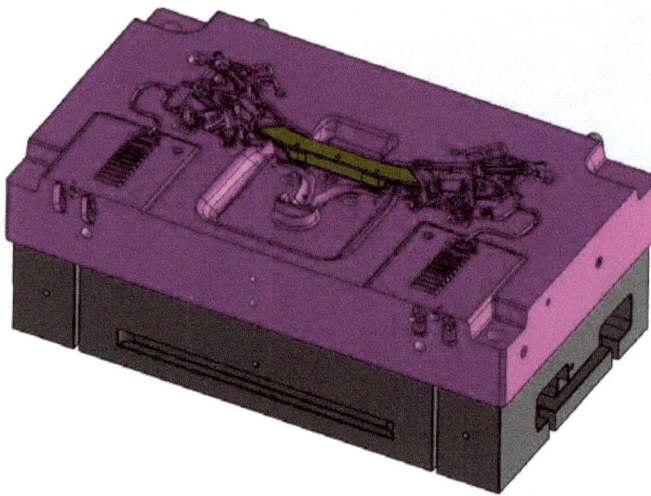

(b)

Figure 2. (**a**) 3D model of the closed profile crossbeam with ceramic core. (**b**) The geometry and the positioning solutions in the mold cavity adopted for the ceramic core.

2.1.2. Material Properties

The materials analyzed for the selection of the core's features are Al_2O_3 + SiO_2 + K_2O ceramic samples, characterized by different percentages of these oxides. More details can be found in [49,50]. Three different sets of samples (hereafter named "a", "b", and "c"), with dimensions of $11 \times 11 \times 300$ mm^3 for samples "a" and $6 \times 8 \times 85$ mm^3 for samples "b" and "c", were industrially produced and analyzed. Different sintering temperatures were considered for samples "a", "b", and "c": T = 892, 900, and 908 °C (specimens "a"), and T = 1075 and 1110 °C (specimens "b" and "c", respectively). The properties of these samples will be analyzed as detailed in the following section. After that, the hollowed aluminum high pressure die casting crossbeam prototypes made of EN AC-43500 aluminum were produced. Table 1 shows the chemical composition ranges for the alloy used.

Table 1. Chemical composition of EN AC-43500 aluminum alloy.

Chemical Elements	Si	Fe	Cu	Mn	Mg	Ni	Zn	Sn	Ti
EN AC-43500	9.5–11.5	0.15	0.03	0.5–0.8	0.1–0.5	-	0.07	-	0.15

2.2. Experimental and Numerical Setups

2.2.1. FEA

Structural Simulation

A necessary premise for this section is that the evaluation of the original structural performances is only feasible when the real crossbeam load conditions are known, and these depend on several factors, such as type of car, frame position, bounded component loads, and standard vehicle use (straight road, curves, etc.). These loading conditions are confidential data and, therefore, cannot be disclosed; for this reason, this experimental research was developed considering dummy loads, based on field experience.

Notwithstanding this limit, it is important to underline that careful comparative work between the original structural solution and the new modified one was performed. Due to the complex geometry of the component, an analytical study can be implemented only for estimating the real mechanical behavior and for a preliminary sizing of the boxed solution geometry.

The numerical analyses were carried out through Autodesk Simulation®software. The finite element models employed for both geometries are composed of parabolic brick elements. The finite element model for the original geometry is composed of about 1.4×10^6 degrees of freedom, 2.9×10^5 elements, and 4.5×10^5 nodes, whereas the finite element model for the modified geometry is constituted of about 1.6×10^6 degrees of freedom, 3.1×10^5 elements, and 4.7×10^5 nodes. All analyses were performed considering the material in linear elastic field. Several component simulations were conducted both in the original and in the boxed configuration to compare their mechanical behavior.

The aim of the first numerical analyses is to evaluate the dynamic performance of the components. This element is manufactured for a specific vehicle and it is very important to know the values of the first natural frequencies and relative vibration modes. Moreover, as mentioned above, the stiffness of the boundary conditions is unknown, so two different simulations have been carried out to fill this gap, the first with free-free boundary conditions, and the second with rigid constraints in the component fixing points. These two boundary hypotheses constitute the opposite extreme fixing conditions of the crossbeam during its lifespan. In addition to the modal studies, buckling analyses were performed to evaluate and compare the load capacity of the two different types of components [51–53].

The analyses were conducted with two different external loads. In the first load condition, two forces (each with a value of 100,000 N) were applied to the component, whereas, in the second load condition, two moments (each with a value of 10,000 N·m) were imposed with the same direction of the main axis of the component. The selected load conditions allow for comparison of the mechanical behavior of the two different crossbeam geometries. The magnitude, type, and direction of the external load were carefully selected, based on previous experience, in order to perform a comparative analysis with different load buckling factor values. Figure 3 shows the buckling analysis model considered, while Table 2 indicates the FEM simulation parameters used.

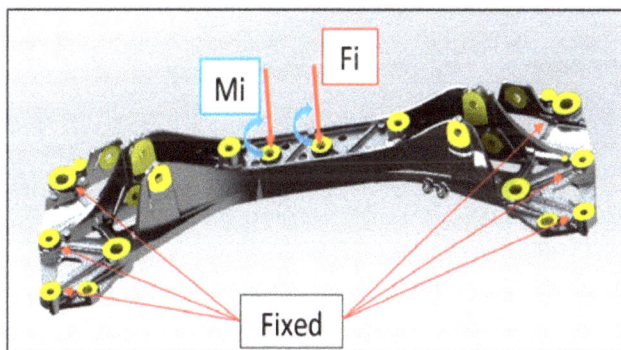

Figure 3. Buckling analysis model: boundary condition and load cases.

Table 2. FEM simulation parameters.

E (Young Modulus) (N/mm²)	ν (Poisson Coefficient)	G (Tangential Modulus) (N/mm²)	ρ (Density) (kg/dm³)
70,000	0.3	30,000	2.7

Process Simulation

Computer aided engineering (CAE) HPDC process simulations were used to forecast and improve the quality of the casting and verify foundry feasibility. In particular, these simulations were useful in the design phase to prevent foundry defect formation (shrinkage porosity, air entrainment, etc.) by changes in the component shape, to guide the die design, leading to the final set of process boundary conditions.

These analyses were carried out with CastleBody from PiQ2 [54,55]. This is a new generation dual phase casting simulation software that can handle both fluids and the spraying phenomena, thanks to a compressible–incompressible dual phase volume of fluid (VOF) formulation. In most simulation software, air is not considered as a moving fluid but only as a steady (not moving) computational domain, and pressure is calculated according to a volume ratio that changes during filling, but no outflow can be modeled. This is a relevant item since, in a foundry process, the filling of a die usually involves two fluids: the molten metal entering into the cavity and the air that is displaced and must exit from it. Their interaction should be carefully analyzed since the pressure increase that develops while filling in the empty mold space due to gas compression leads to partial venting of the air, from the die, through air vents and outflows. Consequently, these analyses are also useful in order to comprehensively evaluate the way air outlets are designed and located.

From the solidification point of view, a potential problem in the production of HPDC castings with disposable ceramic cores could be their different thermal properties compared to the hot work tool steel W1.2343, usually used for the other parts of the die. This issue was faced by modeling the actual thermal properties (temperature, conductivity, and specific heat) of the ceramic core during both filling and solidification calculations, in order to take into account this difference in terms of both flow and shrinkage porosity prediction. Due to its low conductivity, neglecting the heating of the core during the first seconds after it is placed into the mold, the initial ceramic insert temperature was set to the preheating value of 100 °C. A uniform steel temperature of 230 °C was set for the die according to the average value measured on the actual die. The pouring temperature of the liquid alloy was set at 690 °C. The thermal properties of the materials used in the simulation are shown in Table 3, while the the CFD simulation parameters are shown in Table 4.

Table 3. Thermal properties of mold materials.

Ceramic insert			
Thermal Conductivity (W/m/K)	Specific Heat ($m^2/s^2/K$)	Density (kg/m^3)	Initial Temperature (°C)
0.336	818.3	1920	100
H11 W1.2343 Steel Die			
Thermal Conductivity (W/m/K)	Specific Heat ($m^2/s^2/K$)	Density (kg/m^3)	Initial Temperature (°C)
28.6	460	7780	230

Table 4. CFD simulation parameters.

Pouring temperature of liquid alloy	690 °C
Ceramic insert preheating temperature	100 °C
Initial temperature of the die	230 °C
Heat transfer coefficient between liquid Al and mold during filling	4000 W/m^2/K
Heat transfer coefficient between liquid Al and mold during solidification	1200 W/m^2/K
Slow shot plunger speed	0.18 m/s
Fast shot plunger speed	3.5 m/s
Third phase intensified pressure on metal	90 MPa
Shotsleeve active length	0.860 m
Fast shot start stroke	0.563 m

The heat transfer coefficient between the alloy and steel during filling was set at 4000 W/m^{2}K. Both standard and ceramic cored versions of the casting were simulated in order to understand the difference in filling, improve the runners and gating layout for both configurations, and check for filling- or solidification-related defect formation.

According to well-known die design and process parameters calculation guidelines [56], the gate size was calculated so as to avoid excessive metal velocities (>40 m/s) at the gate, in order to reduce the risk of ceramic insert erosion or die soldering while allowing a fast-enough filling time to avoid filling-related defects. Optimal injection parameters (plunger strokes and velocities) for the simulation were set according to theoretical calculations on filling to achieve a suitable filling time of about 75 ms, typical for the average 3.5–5 mm thickness range of the part [56]. According to that, a first phase slow shot speed of 0.18 m/s and a second phase fast shot speed of 3.5 m/s were adopted. The commutation between slow and fast shot was set so that fast shot velocity would be fully developed when metal reached the gate. A final third phase intensified pressure on the metal by 90 MPa, typical for not-crash-relevant structural parts imposed on the solidifying alloy to eliminate or reduce shrinkage defects. Simulations were run on a prevalent hexahedral conformal mesh of about 1,000,000 cells. Local mesh refinement near high speed flow sensitive regions (runners and overflow gatings, thinner sections) was adopted in order to achieve a better flow pattern representation, as shown in Figure 4.

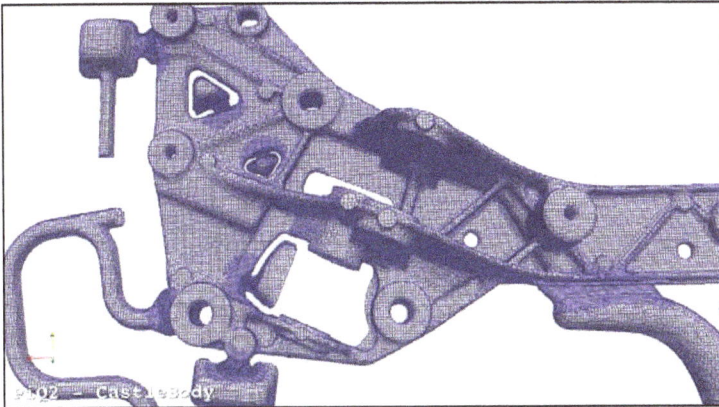

Figure 4. Local mesh refinement.

2.2.2. Material Characterization

Ceramic Samples

The selected ceramic cores were experimentally characterized in terms of:

- Density

For each material, cubic samples having a volume of about 1 mm^3 were cut from the industrially produced bars described in the "Materials and Methods" section. A pycnometer carefully filled with ethanol absolute at ambient temperature was used, and the density was calculated according to Equation (1):

$$\rho_{ceramic} = \rho_{ethanol} \cdot \frac{(m_1)}{(m_1 + m_2 - m_3)},$$ (1)

where

$\rho_{ethanol} = 0.79$ (g/mL);
m_1 = dry sample weight (g);
m_2 = pycnometer fill with ethanol weight (g);

m_3 = pycnometer fill with ethanol + samples weight (g).

- Ceramic Decoring

High pressure water jet technique and different leaching tests were conducted in order to select the proper method for the removal of the ceramic core from the aluminum component. In particular, this latest activity started with the evaluation of the effect of different acids on both ceramic and aluminum samples. Indeed, the leaching method has to be effective only on the core, maintaining an unchanged metal structure. For each selected material, ceramic cubic samples having a volume of about 1 mm^3 were cut from the industrially produced bars. Aluminum samples were considered likewise. Samples were pickled for 1 hour in different distilled water solutions with the following substances: 5% and 10% HF; 65% HNO$_3$; 50% C$_6$H$_8$O$_7$; 50% CH$_3$COOH; 50% and 100% C$_6$H$_8$O$_7$ + CH$_3$COOH (1:1). Finally, the specimens were rinsed in water and dried. The specimens were weighed before and after the tests and the mass loss was determined.

- Three-Point Bending Test

The bending test is one of the most common methods to study the mechanical behavior of brittle ceramic. A three-point bending test was used to determine the flexural strength σ_f (MPa) and the Young Modulus E (GPa), according to UNI EN 843:1 [57] and UNI EN 843:3 [58], respectively. The samples were first dried in a laboratory oven at T = 60 °C for 30 min in order to remove potential moisture that can affect mechanical properties. The experimental setup includes the bending device and an electromechanical Instron 3369 testing machine with a 50 kN loading cell. A constant crosshead displacement speed of 0.5 mm/min was employed. The pins' diameter was 10 mm, and their span "L" was 80 mm for material "a" and 60 mm for materials "b" and "c", selected on the basis of specimen dimension. The flexural strength σf (MPa) was calculated according to Equation (2) [57]:

$$\sigma_f[MPa] = \frac{3F_{max}L}{2bh^2} \tag{2}$$

where

F_{max} [N] = maximum load;
b [mm] = sample width, corresponding to the side of the bar orthogonal to the direction of the load;
h [mm] = sample thickness.

The test for the calculation of the Young modulus (according to Equation (3) [58]) consists in six loading and unloading cycles from 0 N to a load $F \leq F_r$, where F_r is the load at break obtained from the flexural strength tests. The same cycles were replicated on a reference steel bar sized 15 × 11 × 130 mm^3.

$$E_i\ [MPa] = \frac{(F_2 - F_1)L^3}{4bh^3(d_c - d_s)} \tag{3}$$

where

F_1 [N] = 10% F_{max};
F_2 [N] = 90% F_{max};
L [mm] = pin span;
b [mm] = sample width;
h [mm] = sample thickness;
d_c [mm] = sample displacement, in the range between F_1 and F_2;
d_s [mm] = reference steel bar displacement, in the range between F_1 and F_2.
Hollowed Aluminum High Pressure Die Casting Component

The prototypes were experimentally characterized in terms of:

- Microstructure

Transverse sections of the hollowed component in the as-cast condition were obtained for microstructural characterization. They were wet-ground through successive grades of SiC abrasive papers from P120 to P1200, followed by diamond finishing to 0.1 μm. The samples were examined using optical microscopy (OM) using a Leica DMI 5000M (Leica Microsystem, Milan, Italy) and scanning electron microscopy (SEM) using a LEO EVO 40 (Zeiss, Milan, Italy). Semiquantitative chemical analyses were obtained by means of an Energy Dispersive Spectroscopy (EDS)–Link Analytical eXL probe (Oxford Instruments, Milan, Italy), with a spatial resolution of a few microns.

-   Hardness

Vickers microhardness (HV) tests were carried out on a transverse section of the hollowed component under 2.94 N (0.3 kgf) load applied for 15 s, by means of a Micro Duromat 4000 Reichert Jung instrument, according to ASTM E92-16 and ASTM E140-02. The most proper method to assess the resistance of this component is micr-hardness. Indeed, tensile specimens cannot be machined from the component due to its geometry. In addition, the hardness profile would provide information about the overall mechanical properties along the entire cross-section, highlighting eventual local instability. These hardness gradients could be related both to the potential presence of defects typical of casting (i.e., porosities), and to possible effects at the interface between the ceramic core and the aluminum component.

3. Results and Discussion

3.1. FEA Results

3.1.1. Mechanical Behaviour Simulation

The results of the component displacement with respect to the two different boundary conditions applied in the finite element model (free-free and fixed) are reported in Figures 5 and 6.

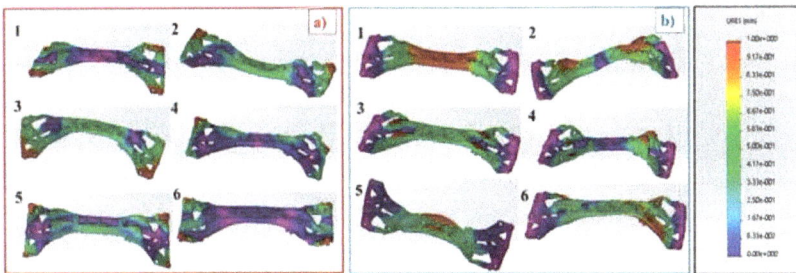

Figure 5. Displacement trend for the first six vibration modes for original geometry: (**a**) free-free boundary conditions and (**b**) fixed conditions.

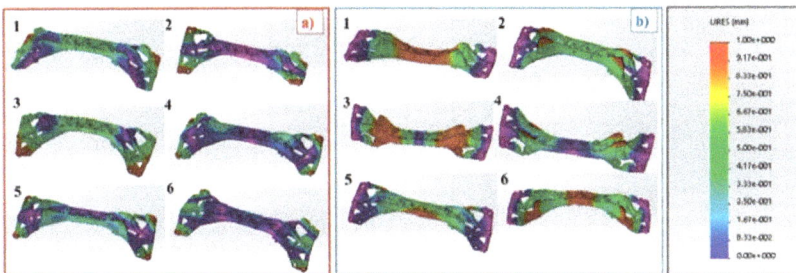

Figure 6. Displacement trend for the first six vibration modes for new geometry: (**a**) free-free boundary conditions and (**b**) fixed conditions.

In particular, Figure 5 refers to the original HPDC geometry, while Figure 6 shows the new one, planned for HPDC with ceramic core application. As is known, the displacement of a structure is a function of the specific vibration mode value; in these simulation results, the component area subjected to the maximum deformability and its consequent flexibility are highlighted. In particular, the red and blue colors represent the maximum and minimum crossbeam displacement, respectively. The values of the first six natural frequencies are reported in Table 5.

Table 5. Natural frequencies values.

First Six Natural Frequencies	Original Geometry		New Geometry	
N°	Free-Free (Hz)	Fixed (Hz)	Free-Free (Hz)	Fixed (Hz)
1	226.2	251.1	330.1	264.5
2	278.1	603.2	349.2	625.3
3	412.1	620.3	408.2	653.5
4	632.6	810.1	677.3	905.6
5	733.3	1044.4	745.5	1128.1
6	882.4	1250.4	899.3	1282.9

The modal analysis results pointed out that the frequencies of the new geometry were higher than the original one, both for free-free and fixed boundary conditions. In general, the boxed solution exhibits an increase of the natural frequencies between 10% and 15%, compared to the values of the original solution. The greatest increase of the natural frequencies is provided by the modes involving the component's central area, with particular relevance under the torsional modes. The value of torsional frequency for the free-free boundary condition changed from 226.2 Hz for the original geometry to 349.2 Hz for the boxed solution, with an increase of about 54.3%. This is an excellent result for this type of component because the first natural frequency values are among the most important design parameters. They have to be over a specific value which depends on vehicle type [59,60].

It is also important to underline that the increase of the component stiffness with the natural frequencies determines also an enhancement of the whole chassis stiffness and, therefore, as is known, vehicle drivability improves, especially on a curve. As regards the extremities of the crossbeam, i.e., the fastening zone, the dynamic behavior of the two components is very similar because this zone is substantially unchanged, due to the interchangeability required by the new component design. The results concerning the buckling analyses are reported in Figures 7 and 8; in particular, Figure 7 concerns the original HPDC geometry, while Figure 8 refers to the new one, planned for HPDC with ceramic core application. It is worth noting the component's displacement in correspondence to the specific buckling value coefficient for a load case 1, i.e., only two forces applied to the component; and b lode case 2, only two moments applied to the component. In Figures 7 and 8, the red and blue component's areas represented the maximum and minimum displacement, respectively.

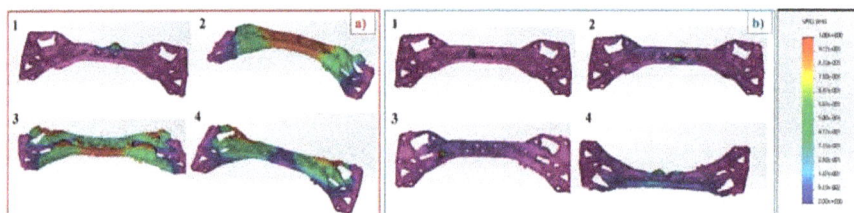

Figure 7. Displacement trend for the first four deformations for original geometry: (**a**) load case 1 (F); (**b**) load case 2 (M).

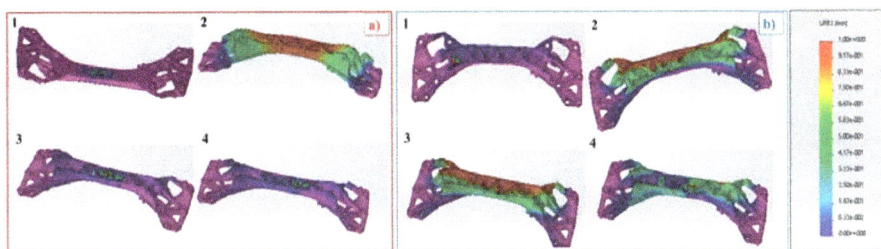

Figure 8. Displacement trend for the first four deformations for new geometry: (**a**) load case 1 (F); (**b**) load case 2 (M).

Finally, the buckling coefficients are reported in Table 6; these values multiplied by the nominal load (F or M for load case 1 and 2, respectively) represent the load for the elastic buckling phenomena of the structure. From the numerical buckling analyses of both the force and moment applied (corresponding to the supposed loading conditions during the component's lifespan), it is clear that these values are significantly higher in the new solution compared with the original one. This positive result is valid also in the case where the buckling factors values are negative, i.e. when an external load is applied in the opposite direction. This result implies that it is possible to increase the magnitude of the load applied to the component avoiding the buckling's failure; which means that, under load equal condition, the new solution has a bigger safety factor than the original solution. In addition, in the new geometry, it should be noted that the areas close to the holes—implemented to guarantee the alignment of the core in the central area—show the maximum deformability.

Table 6. Load buckling factors for two geometries and for two different external loads.

Buckling Coefficients	Original Geometry		New Geometry	
N°	Buckling Factors for Load Case 1 (F)	Buckling Factors for Load Case 2 (M)	Buckling Factors for Load Case 1 (F)	Buckling Factors for Load Case 2 (M)
1	−1.415	−0.881	−1.515	−1.539
2	1.107	1.200	1.508	1.565
3	1.742	1.254	1.706	1.931
4	1.832	1.358	1.951	1.991

This implies that the new component's geometry can be considerably improved by acting on limited areas and, in particular, the zone around the holes for the core fixing. Overall, it is possible to affirm that the new geometry, planned for HPDC with ceramic core application, has a much higher performance than the original HPDC solution as regards both the dynamic behavior (correlated to the natural frequencies values) and the external loads (correlated to the buckling values).

3.1.2. Process Simulation

Standard without-core and ceramic-cored HPDC versions of the casting were simulated in order to understand the difference in filling, improve the runner and gating layout, and to assess core erosion risks and filling- or solidification-related defect formation. Since integrity is mandatory for this part and entrained air bubbles can reduce tensile and fatigue properties, air entrapment was analyzed first.

The filling analysis of both standard and cored versions of the casting shows that only residual amounts of air can be detected in the parts. A threshold visualization was used in order to compare the behavior of the two geometries at the end of filling (Figures 9–11). This is a standard procedure for the analysis of casting simulation. Figure 9a,b shows in blue those mesh cells containing more than 1% and 3% of air in volume respectively. The results show that only a few cells in the part contain entrapped air. In addition, those cells are mainly located in the dead ends of the casting and, therefore, air can be easily evacuated through a further re-design of the overflows. The situation is very similar in both versions of the casting, but the cored one presents somehow a low increase of air entrapment in the center of the part, just between the two gatings. Also, these points could be improved by evacuating air from the cavity through chill vents or vacuum applications.

Figure 10a,b shows the temperature distribution of the alloy at the end of filling in the range between liquidus and solidus temperatures. Some regions of the casting are filled by partially solidified alloy that could potentially lead to the formation of cold joints.

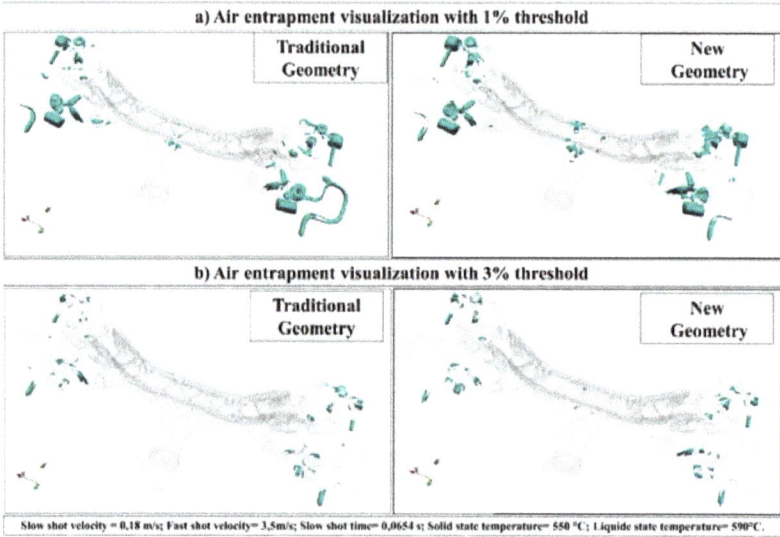

Slow shot velocity = 0,18 m/s; Fast shot velocity= 3,5m/s; Slow shot time= 0,0654 s; Solid state temperature= 550 °C; Liquide state temperature= 590°C.

Figure 9. Air entrapment visualization (**a**) with 1% threshold and (**b**) with 3% threshold for traditional and new geometry.

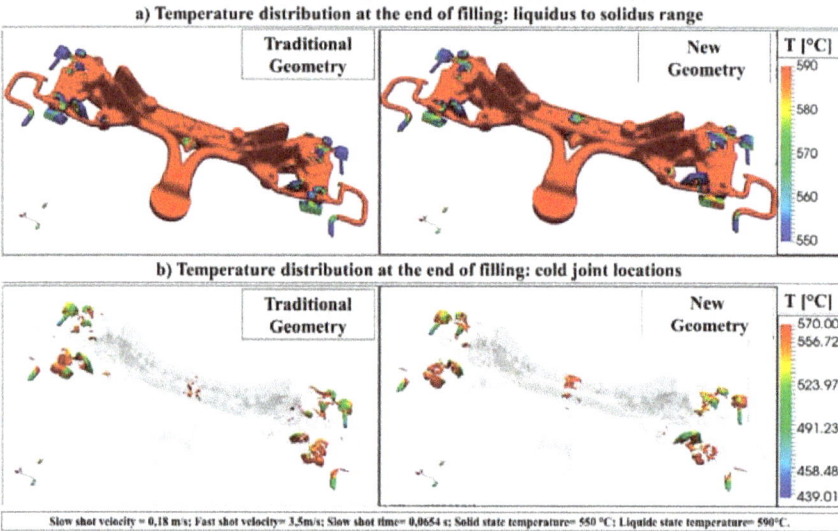

Slow shot velocity = 0,18 m/s; Fast shot velocity= 3,5m/s; Slow shot time= 0,0654 s; Solid state temperature= 550 °C; Liquide state temperature= 590°C.

Figure 10. (**a**) Temperature distribution at the end of filling: liquidus to solidus range; (**b**) Temperature distribution at the end of filling: cold joint locations.

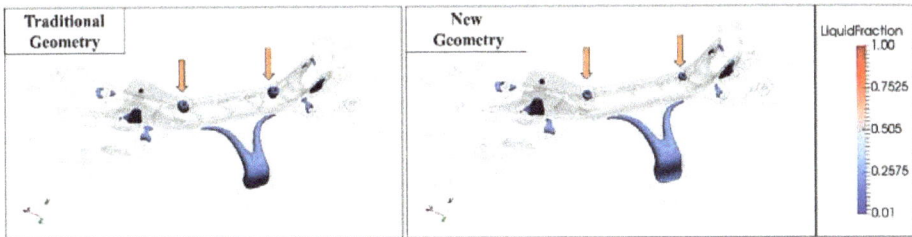

Figure 11. Results of solidification analysis.

In both versions of the casting, most of the potential defects are located in the fixing embossments, far from the ceramic core, which could be an issue to take into account during the assembly of the component. It is worthwhile to note that air entrapment is very similar in both cored and traditional castings, except for the central embossment of the cored one, which shows a pocket of cold metal that should be avoided by improving the process and carrying out a final review of the design. These expedients will allow for further increasing of the soundness of these regions of the casting, which is only slightly affected by the ceramic core introduction into the process.

The solidification analysis (Figure 11) shows some liquid alloy hotspots at the end of the filling that could generate some shrinkage porosity. The results show a slight improvement of the new geometry behavior (cf. the arrows in Figure 11) in comparison with the traditional one. Indeed, the potential risk of shrinkage porosity in the two embossments located on the sides of the ceramic core is lower (smaller liquid fraction pockets) in the new solution, probably due to the reduction in thickness of the casting near this area, resulting in a faster solidification. This could be further improved in both solutions with the introduction of squeeze pins into the die.

The filling simulation process lasts about 0.2 s, whereas the solidification process lasts 8.4 s. These times are in line with traditional HPDC production of components of a similar size. The process simulation was useful to highlight some process and tools improvement (i.e., die, local cooling, squeeze, chill vents, gate injection, etc.). It is worthwhile to note that the defects highlighted in the innovative geometry are very similar to the defects found in the standard geometry.

Finally, the results allow us to affirm that the new HPDC with ceramic core solution is feasible and able to produce complex components with a quality very similar to the traditional process.

3.2. Material Characterization Results

3.2.1. Ceramic Samples

Density

The density of the ceramic samples is 1.98 g/cm^3 for sample "a" and 1.92 g/cm^3 for sample "b" and "c". No relevant differences were observed between the different materials tested.

Ceramic Decoring

After preliminary tests, the different sintering temperatures of the ceramic samples have proven not to be relevant parameters. The effectiveness (indicated with ✓) or ineffectiveness (indicated with ✗) of the leaching on ceramic and the effect observed on aluminum samples are reported in Table 7. In particular, these tests pointed out that ceramic samples are etched only by HF solutions, which have a well-known detrimental effect on aluminum. Indeed, aluminum samples burned after the treatment in HF solution. Considering the unpromising leaching results, core removal was carried out on the prototypes using a high pressure water jet. This last operation successfully removed the core and maintained at the same time the component's integrity. These observations are documented by the microscopic analysis reported in the following section.

Table 7. Ceramic decoring effectiveness (✓) or ineffectiveness (✗).

Type of Ceramic Decoring Tested	HF		HNO$_3$		C$_6$H$_8$O$_7$	CH$_3$COOH	C$_6$H$_8$O$_7$/CH$_3$COOH (1:1)	
Acid amount (water solution)	5%	10%	65%		50%	50%	50%	100%
Immersion with ultrasound	no	no	no	yes	yes	yes	yes	yes
a	✓	✓	✗	✗	✗	✗	✗	✗
b	N/A	N/A	N/A	N/A	✗	✗	✗	✗
c	N/A	N/A	N/A	N/A	✗	✗	✗	✗
Aluminum	Burned-out		No effect		No effect	No effect	No effect	

- Three-point bending test

Table 8 reports the average values and standard deviation of Young modulus E (GPa) and flexural strength σ_f (MPa) calculated after the three-point bending test on the different ceramic samples. Table 6 shows that both E and σ_f increased as the sintering temperature increased; this effect is particularly remarkable in sample "a". The best match in terms of mechanical properties according to Moschini et al. [50] was found in samples "c-1110 °C", i.e., in the material selected for the production of the hollowed aluminum component.

Table 8. Average values and standard deviation of the Young modulus E (GPa) and flexural strength σ_f (MPa).

Samples	Sintering T (°C)	E (GPa)		σ_f (MPa)	
		Avg.	±	Avg.	±
a	892	13.26	0.96	14.13	1.7
	900	13.59	1.72	14.07	1.43
	908	15.45	0.83	16.28	1.91
b	1075	9.2	0.76	13.02	0.92
	1110	10.14	0.33	13.41	0.52
c	1075	12.12	1.35	17.56	1.82
	1110	13.97	0.97	19.41	0.56

3.2.2. Hollowed Aluminum High Pressure Die Casting Component

Some hollowed aluminum high pressure die casting prototypes (about 100 pieces) were produced on an IDRA OLS2000 HPDC machine (IDRA group, Brescia, Italy) equipped with automatic ladle, 6-axis robot for ejection and die lubricant spraying. In particular, the prototypes are manufactured using the same casting conditions applied for the process simulations, while the cycle time was approximately 70 s. Next, the ceramic core was removed with a water jet. The crossbeam prototype is shown in Figure 12.

Microstructure

Figure 13 shows one of the transverse sections of a hollowed component in the as-cast condition observed with the optical microscope. In particular, each image is an overview of the entire component's thickness along all its sides, composed by a collage of various micrographs. Figures 14 and 15 show some details at higher magnifications obtained with OM and with SEM equipped with EDS, respectively.

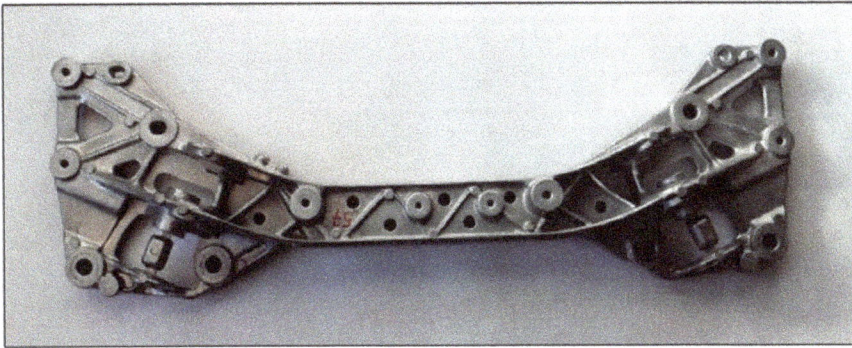

Figure 12. Hollowed aluminum high pressure die casting component.

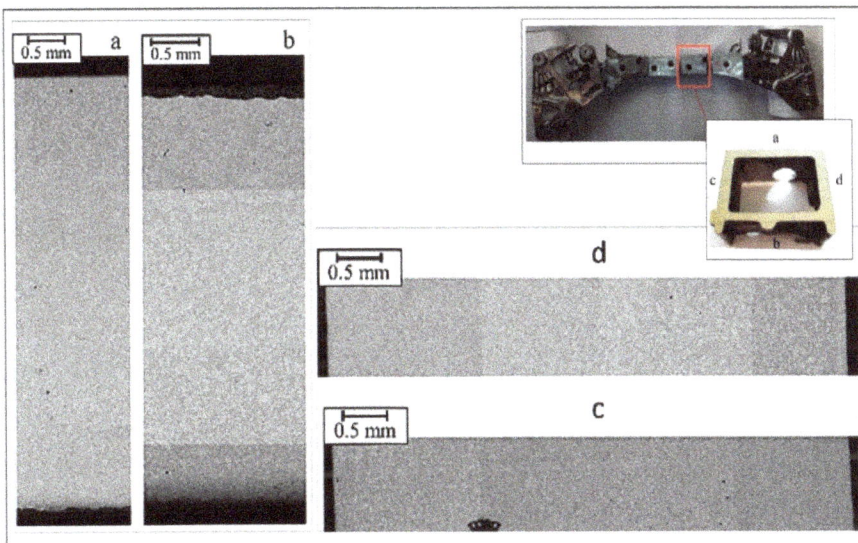

Figure 13. Collage of various micrographs of the transverse sections of the hollowed crossbeam.

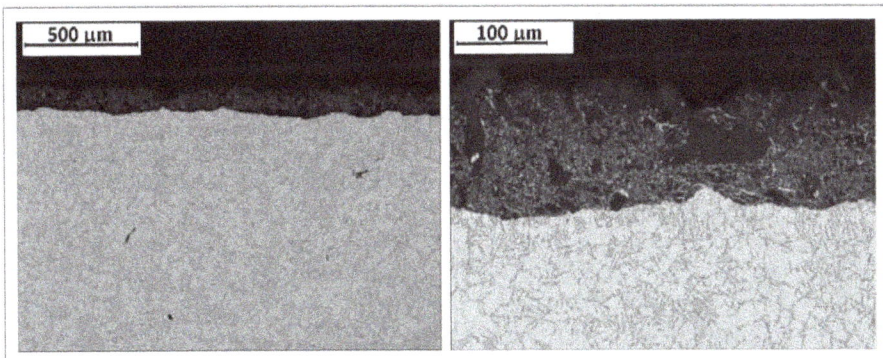

Figure 14. Optical microscope details of the transverse sections of the hollowed crossbeam.

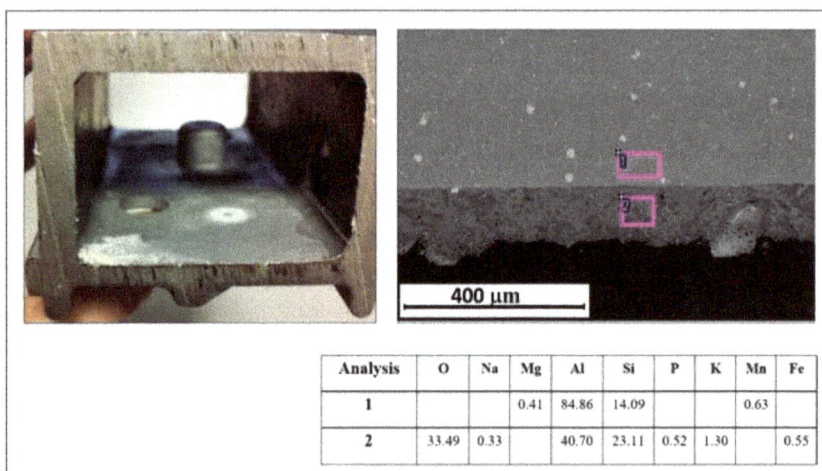

Analysis	O	Na	Mg	Al	Si	P	K	Mn	Fe
1			0.41	84.86	14.09			0.63	
2	33.49	0.33		40.70	23.11	0.52	1.30		0.55

Figure 15. Visual inspection on cermet and SEM image with results of EDS analysis (wt%).

The OM analyses on the as-cast samples show, for all the samples, a typical microstructure composed of α-Al matrix with Al-Si eutectic in the interdendritic space and some intermetallic compounds. In particular, precipitates in EN AC-43500 are usually composed of α-Al_{15}(Mn,Fe)$_3$Si$_2$ polyhedral structure [61]. In addition, a cermet compound of about 100 μm thickness was observed in some areas of the samples at the interface between aluminum and ceramic core.

Figure 15 shows an example of backscattered SEM appearance of the interface layer between aluminum alloy and cermet. The limits of the instrument with light elements like oxygen are known; therefore, the EDS analysis reported in the figure does not reflect the absolute oxygen amount in the cermet oxides or in the interface, but it is a semiquantitative chemical analysis. The high amount of oxygen observed on the surface of the cavity does not represent the exact value; however, considering also the presence of Al, Si, and K, it certainly indicates the residual presence of cermet.

This cermet is a ceramic–metal mixture that exhibits high strength and is ideally designed to have the optimal properties of both ceramic, such as high temperature resistance and hardness, and metal, such as the ability to undergo plastic deformation [62–64]. In this work, according to Molina et al. [49], the cermet layer results coherent with aluminum surface and shows no tendency to crush. To prevent this phenomena, a specific coating can be applied on the ceramic cores. It is important to highlight that the cermet has no consequence for the studied application.

Vickers Hardness HV

Figure 16 shows the variation of the Vickers microhardness HV0.3 as a function of the thickness along the orthogonal section of the component of Figure 13. The position of the cross-section profiles from "a" to "d" is reported in Figure 13. High variability in the hardness values along the thickness (especially for the "a" and "d" areas of Figure 13) is noted, probably due to the presence of defects that affect the measurement. Indeed, these values were only considered to assess the alloy's overall mechanical properties. These defects are typical of HPDC process and are not affected by the ceramic core introduction; only the different values of hardness observed at the interface between the ceramic core and the aluminum component are related to the cermet layer. It should be noted that it was not always possible to measure the hardness in correspondence of this composite due to its particular structure. The average overall microHV is comprised in a range between 75 and 90 HV, which is typical of similar actual case studies that used the same alloy with the traditional HPDC technology (without ceramic core) [5]. This confirmed that the slight extension of the cermet layer does not affect the microstructure and the mechanical properties in the remaining component section.

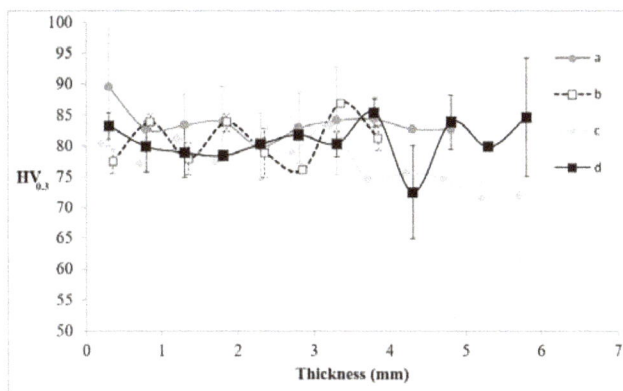

Figure 16. Vickers microhardness HV0.3 variation with respect to the thickness along the transverse section of Figure 13.

4. Conclusions

This paper describes the study, the development, and the realization of a new HPDC method with the use of a new ceramic core that allows the production of a lighter and improved safety-relevant automotive component.

In particular, this research investigated the production of an improved aluminum crossbeam for passenger cars with HPDC and innovative lost ceramic cores. The main results in terms of process, structural simulations, and experimental tests can be summarized as follows:

- Despite the thinness of the original component, a slight decrease in weight was achieved thanks to the design modifications applied to the new boxed part.
- The modal analysis results pointed out that the frequencies of the new geometry were higher than those of the original geometry. From the numerical buckling analyses, it is clear that these values are significantly higher in the new solution compared with the original one. Therefore, the new geometry has much higher performance than the original solution, as regards both the dynamic behavior of the component and the external loads that act on the component through the vehicle's chassis.
- The process simulations confirmed that the new HPDC with ceramic core solution is feasible and able to produce complex components with a quality very similar to the traditional process.
- The results of the ceramic experimental characterization allowed for selection of the best match of mechanical properties for the ceramic cores. Approximately 100 hollowed crossbeam prototypes were produced, and the as-cast condition was characterized by means of metallurgical analyses and hardness tests. The results confirm that the actual case study has properties and a microstructure very similar to the component produced with the traditional HPDC technology (without ceramic core). A trace of cermet compound was observed in some areas at the interface between aluminum and ceramic core which is not an issue in this component.

The obtained results demonstrates that HPDC with ceramic lost cores has excellent potential for high production volumes and near-net shape components, with the possibility to produce hollow, thin, and complex parts, contributing to making HPDC competitive with respect to the conventional casting process for the production of automotive hollow parts.

This experimental work has opened the way to future developments that range from the possibility to improve crossbeam lightweighting to the implementation of this technology for other automotive hollow components, such as pillar or different parts of the vehicle chassis.

Author Contributions: Concept development (provided idea for the research): G.C., D.D., M.F., C.G., L.S., A.P., S.C.; Design (planned the methods to generate the results): G.C., D.D., M.F., A.P., S.C.; Supervision (provided oversight, responsible for organization and implementation): M.F., C.G., A.P.; Data collection/processing (responsible for experiments, patient management, organization, or reporting data): G.C., D.D., L.S., A.P., S.C.; Analysis/interpretation (responsible for statistical analysis, evaluation, and presentation of the results): S.C., G.C.; Literature search (performed the literature search): S.C., G.C.; Writing (responsible for writing a substantive part of the manuscript): S.C., G.C., L.S., A.P.; Critical review (revised manuscript for intellectual content, this does not relate to spelling and grammar checking): G.C., D.D., M.F., C.G., L.S., A.P., S.C.

Funding: This work was supported by Regione Lombardia—MIUR research program "S.A.V.E." [Grant Number ID 30023402].

Acknowledgments: This paper is dedicated to the memory of Renzo Moschini, who recently passed away. He contributed greatly on the study of ceramic core. This work was partly funded by Regione Lombardia–MIUR (research program "S.A.V.E." ID 30023402). The Authors are grateful to Regione Lombardia, Valentina Ferrari and all the project partners for their collaboration.

Conflicts of Interest: The authors declare no conflict of interest.

References

1. Chindamo, D.; Gadola, M. Reproduction of real-world road profiles on a four-poster rig for indoor vehicle chassis and suspension durability testing. *Adv. Mech. Eng.* **2017**, *9*, 1–10. [CrossRef]

2. Chiaberge, M. *New Trends and Developments in Automotive Industry*; InTech: London, UK, 2011; ISBN 978-953-307-999-8. [CrossRef]

3. Tempelman, E. Multi-Parametric study of the effect of materials substitution on life cycle energy use and waste generation of passenger car structures. *Transp. Res. Part D Transp. Environ.* **2011**, *16*, 479–485. [CrossRef]

4. Cecchel, S.; Chindamo, D.; Turrini, E.; Carnevale, C.; Cornacchia, G.; Gadola, M.; Panvini, A.; Volta, M.; Ferrario, D.; Golimbioschi, R. Impact of reduced mass of light commercial vehicles on fuel consumption, CO2 emissions, air quality, and socio-economic costs. *Sci. Total Environ.* **2018**, *613–614*, 409–417. [CrossRef] [PubMed]

5. Cecchel, S.; Ferrario, D.; Panvini, A.; Cornacchia, G. Lightweight of a cross beam for commercial vehicles: Development, testing and validation. *Mater. Des.* **2018**, *149*, 122–134. [CrossRef]

6. Cecchel, S.; Ferrario, D. Numerical and experimental analysis of a high pressure die casting Aluminum suspension cross beam for light commercial vehicles. *La Metallurgia Italiana* **2016**, *6*, 41–44.

7. Morello, L.; Rosti Rossini, L.; Pia, G.; Tonoli, A. *The Automotive Body: Volume I: Components Design*; Springer: New York, NY, USA, 2001; ISBN 978-94-007-0513-5.

8. Cecchel, S.; Cornacchia, G.; Panvini, A. Cradle-to-Gate Impact Assessment of a High-Pressure Die-Casting Safety-Relevant Automotive Component. *JOM* **2016**, *8*, 2443–2448. [CrossRef]

9. Cecchel, S.; Chindamo, D.; Collotta, M.; Cornacchia, G.; Panvini, A.; Tomasoni, G.; Gadola, M. Lightweighting in light commercial vehicles: Cradle-to-grave life cycle assessment of a safety relevant component. *Int. J. Life Cycle Assess.* **2018**, *23*, 1–12. [CrossRef]

10. Cecchel, S.; Collotta, M.; Cornacchia, G.; Panvini, A.; Tomasoni, G. A comparative cradle-to gate impact assessment: Primary and secondary aluminum automotive components case. *La Metallurgia Italiana* **2018**, *2*, 46–55.

11. Cecchel, S.; Cornacchia, G.; Gelfi, M. Corrosion behavior of primary and secondary AlSi High Pressure Die Casting alloys. *Mater. Corros.* **2017**, *68*, 961–969. [CrossRef]

12. Solazzi, L. Applied research for Weight Reduction of an industrial Trailer. *FME Trans.* **2012**, *40*, 57–62.

13. Solazzi, L. Wheel rims for industrial vehicles: Comparative and experimental analyses. *Int. J. Heavy Veh. Syst.* **2011**, *18*, 214–225. [CrossRef]

14. Hirsch, J. Aluminium in Innovative Light-Weight Car Design. *Mater. Trans.* **2011**, *52*, 818–824. [CrossRef]

15. Chindamo, D.; Lenzo, B.; Gadola, M. On the vehicle sideslip angle estimation: A literature review of methods, models and innovations. *Appl. Sci.* **2018**, *8*, 355. [CrossRef]

16. Henriksson, F.; Johansen, K. On Material Substitution in Automotive BIWs – From Steel to Aluminum Body Sides. *Procedia CIRP* **2016**, *50*, 683–688. [CrossRef]

17. Kelkar, A.; Roth, R.; Clark, J. Automobile Bodies: Can Aluminum Be an Economical Alternative To Steel? *JOM* **2001**, *53*, 28–32.

18. Zhoua, J.; Wana, X.; Lia, Y. Advanced aluminium products and manufacturing technologies applied on vehicles at the EuroCarBody conference. *Mater. Proc.* **2015**, *2*, 5015–5022. [CrossRef]

19. Fridlyander, I.N.; Sister, V.G.; Grushko, O.E.; Berstenev, V.V.; Sheveleva, L.M.; Ivanova, L.A. Aluminum alloys: Promising materials in the automotive industry. *Met. Sci. Heat Treat.* **2002**, *44*, 365–370. [CrossRef]

20. Dioni, D.; Cecchel, S.; Cornacchia, G.; Faccoli, M.; Panvini, A. Effects of artificial aging conditions on mechanical properties of gravity cast B356 aluminum alloy. *Trans. Nonferrous Met. Soc. China* **2015**, *25*, 1035–1042. [CrossRef]

21. Faccoli, M.; Dioni, D.; Cecchel, S.; Cornacchia, G.; Panvini, A. Optimization of heat treatment of gravity cast Sr-modified B356 aluminum alloy. *Trans. Nonferrous Met. Soc. China* **2017**, *27*, 1698–1706. [CrossRef]

22. Haracopos, B.; Fisher, T.P. *The Technology of Gravity Die Casting*; Hart Pub. Co.: Oxford, UK, 1968.

23. ASM. *Metals Handbook*, 10th ed.; ASM-Metals Park: Geauga, OH, USA, 1990.

24. Campbell, J.; Harding, R.A. *Casting Technology*; TALAT 2.0 [CD-ROM]; EAA: Bruxelles, Belgium, 2000.

25. Schleg, S.; Kamicki, D.P. *Guide to casting and moulding processes. Engineered casting solutions*. Technical Articles, 2000.

26. Brown, J.R. *Foseco Non-Ferrous Foundryman's Handbook*; Elsevier: Oxford, UK, 1999; ISBN 9780080531878.

27. Street, A.C. *The Diecasting Book*, 2nd ed.; Portcullis Press: London, UK, 1990.

28. Nagendra Parashar, S.; Mittal, R.K. *Elements of Manufacturing Processes*; PHI learning Pvt. Ltd.: New Delhi, India, 2006.

29. Vinarcik, E.J. *High Integrity Die Casting Processes*; John Wiley & Sons: Hoboken, NJ, USA, 2002; ISBN 978-0-471-20131-1.

30. Andresen, W. *Die Cast Engineering: A Hydraulic, Thermal, and Mechanical Process*; CRC Press: Boca Raton, FL, USA, 2004.

31. Jelínek, P.; Adámková, E.; Mikšovský, F.; Beňo, J. Advances in technology of soluble cores for die castings. *Arch. Foundry Eng.* **2015**, *15*, 29–34. [CrossRef]

32. Pierri, D. Lost Core: New Perspectives in Die Casting. Available online: https://www.buhlergroup. com/northamerica/en/industry-solutions/die-casting/latest-news/details-7797.htm?title= (accessed on 9 February 2019).

33. Czerwinski, F.; Birsan, G.; Benkel, F.; Kasprzak, W.; Walker, M.J.; Smith, J.; Trinowski, D.; Musalem, I. Developing strong core technology for high pressure die casting. *Automot. Mater.* **2017**, *8*, 1–11.

34. Rupp, S.; Heppes, F. La rivoluzione nella pressofusione. *Tecnico-Industria Fusoria* **2017**, *4*, 72–76.

35. Donahue, R.J.; Degler, M.T. Congruent Melting Salt Alloys for Use as Salt Cores in High Pressure Die Casting. U.S Patent US9527131B1, 25 August 2014.

36. Radadiya, V.A.; Dave, K.G.; Patel, K.R. Design and analysis of salt core for a casting of alluminium alloys. *Int. J. Adv. Eng. Res. Dev.* **2015**, *2*, 344–348.

37. Jelinek, P.; Miksovsky, F.; Beoo, J.; Adamkova, E. Development of foundry cores based on inorganic salts. *MTAEC9* **2013**, *47*, 689–693.

38. Jelínek, P.; Adámková, E. Lost cores for high-pressure die casting. *Arch. Foundry Eng.* **2014**, *14*, 101–104. [CrossRef]

39. Yaokawa, J.; Miura, D.; Anzai, K.; Yamada, Y.; Yoshii, H. Strength of salt core composed of alkali carbonate and alkali chloride mixtures made by casting technique. *Mater. Trans.* **2007**, *8*, 1034–1041. [CrossRef]

40. Yaokawa, J.; Koichi, A.; Yamada, Y.; Yoshii, H. Strength of salt core for die casting. In Proceedings of the International Conference CastExpo '05 NADCA, St. Louis, MO, USA, 16–19 April 2005; NADCA: St. Louis, MO, USA, 2005.

41. Fuchs, B.; Eibisch, H.; Körner, C. Core viability simulation for salt core technology in high-pressure die casting. *Int. J. Metalcast.* **2013**, *7*, 39–45. [CrossRef]

42. Fuchs, B.; Körner, C. Mesh resolution consideration for the viability prediction of lost salt cores in the high pressure die casting process. *Prog. Comput. Fluid Dyn.* **2014**, *14*, 24–30. [CrossRef]

43. Mizukusa, Y. Casting Apparatus and Casting Method for Producing Cylinder Block. U.S. Patent 5,690,159, 27 August 1996.

44. Ackerman, A.D.; Aula, H.A. Method of Making a Cast Aluminum Based Engine Block. U.S. Patent 4,446,906, 8 May 1984.

45. Brown, W.N.; Robinson, P.M. Soluble Metal Casting Cores Comprising a Water Soluble Salt and a Synthetic Resin. U.S. Patent 364,549, 22 July 1969.

46. Gibbons, W.A. Core or Filler of Fusible Material for Hollow Vulcanizable Articles. U.S. Patent 1,523,519, 20 January 1925.
47. Foreman, R.W. Mixture and Method for Preparing Casting Cores and Cores Prepared Thereby. U.S. Patent 4,840,219, 20 June 1989.
48. Sakoda, T. Water Soluble Core for Pressure Die Casting and Process for Making the Same. U.S. Patent 3,963,818, 15 June 1976.
49. Molina, R.; Moschini, R. Production of hollow components in high pressure die casting through the use of ceramic lost cores. In Proceedings of the International Conference High Tech Die Casting 2012, Vicenza, Italy, 9–10 April 2012.
50. Moschini, R.; Calzolaro, A.L. Method for Manufacturing Monolithic Hollow Bodies by Means of a Casting or Injection Moulding Process. Patent International Publication No WO 2011/061593 A1, 26 May 2011.
51. Solazzi, L. Innovative Bolted junction with high ductility for circular tubular element. *J. Constr. Steel Res.* **2015**, *112*, 175–182. [CrossRef]
52. Mats, G.; Larson, F.B. *The Finite Element Method: Theory, Implementation and Applications*; Springer: Berlin, Germany, 2010; ISBN 978-3-642-33286-9.
53. Dimitrios, G. *Pavlou Essentials of Finite Element Method for Mechanical and Industrial Engineers*; Elsevier: Amsterdam, The Netherlands, 2015; ISBN 978-0-12-802386-0.
54. Panvini, A.; Molin, D.; Gislon, C. Dual phase simulation for high pressure die casting: Overview and validation of its capabilities. In Proceedings of the HTDC Conference 2016, AIM, Venice, Italy, 22–23 June 2016.
55. Panvini, A.; Gislon, C. Air entrapment prediction in diecasting through dual phase simulation. In Proceedings of the Die Casting Congress Tabletop 2013, Louisville, KY, USA, 16–18 September 2013.
56. Miller, A. *PQ2 and Gating Manual*; NADCA: Arlington Heights, IL, USA, 2016.
57. UNI EN 843-1:2007. *Advanced Technical Ceramics—Mechanical Properties of Monolithic Ceramics at Room Temperature—Part 1: Determination of Flexural Strength*; NSAI: Dublin, Ireland, 2007.
58. UNI EN 843-2:2007. *Advanced Technical Ceramics—Mechanical Properties of Monolithic Ceramics at Room Temperature—Part 2: Determination of Young's Modulus, Shear Modulus and Poisson's Ratio*; NSAI: Dublin, Ireland, 2007.
59. Smith, J.H. *An Introduction to Modern Vehicle Design*; Elsevier: Oxford, UK, 2002; ISBN 07506 5044 3.
60. Pacejka, H.B. *Tyre and Vehicle Dynamics*, 2nd ed.; Butterworth-Heinemann: Oxford, UK, 2006; ISBN 0-7506-6918-7.
61. Franke, R.; Dragulin, D.; Zovi, A.; Casarotto, F. Progress in ductile aluminium high pressure die casting alloys for the automotive industry. *La Metallurgia Italiana* **2007**, *5*, 21–26.
62. Tinklepaugh, J.R.; James, R. *Cermets*; Reinhold Publishing Corporation: New York, NY, USA, 1960; ASIN B0007E6FO4.
63. Hanaor, D.A.H.; Hu, L.; Kan, W.H.; Proust, G.; Foley, M.; Karaman, I.; Radovic, M. Compressive performance and crack propagation in Al alloy/Ti2AlC composites. *Mater. Sci. Eng.* **2016**, *72*, 247–256. [CrossRef]
64. Bhattacharya, A.K.; Petrovic, J.J. Ductile phase toughening and R-curve behaviour in a B4C-AI cermet. *J. Mater. Sci.* **1992**, *27*, 2205–2210. [CrossRef]

![metals logo] *metals*

MDPI

Article

Multilayered-Sheet Hot Stamping and Application in Electric-Power-Fitting Products

Bin Zhu, Zhoujie Zhu, Yongmin Jin, Kai Wang, Yilin Wang * and Yisheng Zhang

State Key Lab of Materials Processing and Die and Mould Technology, Huazhong University of Science and Technology, Wuhan 430074, Hubei, China; zhubin26@hust.edu.cn (B.Z.); m201670860@hust.edu.cn (Z.Z.); flymings@163.com (Y.J.); wangkai@hust.edu.cn (K.W.); Zhangys@mail.hust.edu.cn (Y.Z.)
* Correspondence: wangyilin@mail.hust.edu.cn; Tel.: +86-027-87544457

Received: 8 January 2019; Accepted: 8 February 2019; Published: 12 February 2019

Abstract: Traditional electric transmission line fittings, which are always manufactured from thick metal slabs, possess the disadvantage of heavy weight. In this study, a new type of electrical-connection-fitting, clevis-clevis component made of high-strength steel is developed to reduce weight, and a new hot-stamping process for multilayered sheets is proposed to manufacture the component efficiently. First, the structure of the new clevis-clevis component is designed, and the corresponding tool is developed. Second, a flat-tool heat transfer experiment is conducted. The influence of the number of layers and contact pressure on the cooling rate of each sheet is investigated. The optimizing number of layers and contact pressure for the multilayered-sheet, hot-stamping process are obtained. The optimal number of layers is two, and the optimal contact pressure is more than 20 MPa. The final microstructure of each sheet is fully martensitic, and the strength is about 1500 MPa. Finally, U-shaped, double-layer-sheet hot stamping is implemented to produce a typical electrical-connection-fitting, clevis-clevis component. The bearing capacity of a four-layered clevis-clevis is tested through numerical and experimental methods. The new connection-fitting clevis-clevis component exhibits a high load capacity of 280 kN. Compared with that of the traditional component, the weight of the new component is reduced by 60%.

Keywords: hot stamping; transmission line fittings; multilayered sheets; contact heat transfer

1. Introduction

Transmission line fittings are important components related to the safe operation of electrical transmission lines [1]. The components are always made of cast or forged steel, such as Q235 and Q345. The drawbacks of the traditional process are high metal consumption and inhomogeneous distribution of temperature and mechanical properties in the thick profile of the formed parts [2]. Given the continuous expansion of electric power systems and the recent increase in line voltage level, lightweight but strong fitting components made of ultra-high-strength steel present a potential application in electrical transmission fittings.

A means to reduce the weight of fitting components is the use of boron steel. Boron steel possesses a yield strength of about 1100 MPa and a tensile strength of more than 1500 MPa after hot stamping and press hardening [3–5]. The advantage of hot-stamped boron steel over cast or forged steel is that the former is heated, formed, and quenched in a shorter time cycle. The thinner sheet steel allows for shorter heating and holding time (4 min) and faster stamping and quenching in cold tools (10 s) [6]. In addition, the homogeneous and excellent property of the formed part may be obtained for a fast cooling-down process. Machining and assembling is needed to produce a component afterwards; however, this is beneficial as it reduces the weight and improves its energy-consuming and mechanical properties. Thus, a new multilayered-sheet, hot-stamping process was proposed, as shown in Figure 1 [7]. In the multilayered-sheet, hot-stamping process, several sheets were stamped simultaneously.

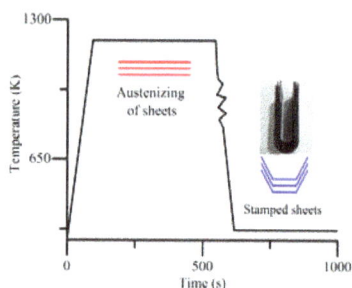

Figure 1. Multilayered-sheet, hot-stamping process.

The key point of the multilayered-sheet, hot-stamping process is to ensure that the cooling rate of each sheet achieves the critical cooling rate (30 K/s) during the quenching process [8], which will determine whether the fully martensitic microstructure will be obtained. Then, the relationship of pressures, number of sheets, and the transmission of heat between the sheet and the tools should be investigated. Yoon indicated the fully martensitic microstructure had been achieved over a very narrow timeline during hot stamping and the quenching process. Moreover, the duration of martensitic expansion is 2.9 s for a low pressure of 2.5 MPa [9]. Seung designed a slice die to improve the cooling rate of the blank during hot stamping and the quenching process [10]. Hay conducted hot stamping tests under different contact pressure values covering the range from 5 to 30 MPa. Temperature measurements in the tool and the blank allowed the estimation of the thermal contact resistance evolution for every contact pressure. Numerical simulation was carried out using a chain which combines thermal and mechanical properties of the hot stamping process [11]. With a similar experimental procedure, Abdulhay investigated the influence of the thermal contact resistance on the contact pressure for both Usibor 1500P and material B [12]. Chao designed a fast-response, temperature measurement and data acquisition system to obtain the temperature history of blank and die under different pressures, and the thermal contact conductance is calculated based on the temperature history data [13]. Applying the corrected thermal contact conductance, the accuracy of the temperature field calculation with the finite element method for hot stamping can been improved. Bai has done similar experimental works with Ti-6Al04V specimens and H13 tool steel. However, in their works, only a single sheet was used in the multilayered-sheet hot stamping as an increase of the number of sheets made the heat transfer between the sheet and the tool more complicated [11,12]. In previous works, hot stamping and a quenching process for a U-shaped sheet were analyzed through numerical and experimental methods. The die gap between the sheet and dies led to diffusional transformation [14]. Controlling the cooling rate of the steel sheet reasonably during stamping and quenching was proposed for obtaining a mixture of multiphase microstructures [15]. The influence of thickness distribution on cooling rate was taken into consideration [16]. In this work, multilayered-sheet, hot-stamping tests are carried out. The number of sheets and the contact pressure are linked to the cooling rate of the sheet. The optimal number of sheets and the contact pressure for the multilayered-sheet, hot-stamping process are also obtained.

The typical electrical-connection-fitting clevis-clevis has a symmetrical U-shape geometry. Thus, the component is suitable for multilayer production. In this study, based on previous works, the clevis-clevis connection was divided into symmetrical layers, and the tool was developed. Then, a multilayered-sheet heat transfer experiment was implemented to confirm the feasibility. Finally, the multilayered-sheet, hot-stamping process was carried out to produce a new structure of clevis-clevis.

2. Features of the New Product and U-Shaped Tool

2.1. Light Weight of Clevis-Clevis Component and Features of the U-Shaped Tool

The characteristics of the new type of clevis-clevis are as follows: a deep, symmetrical, U-shaped structure; the load is borne in one direction; and homogeneity in its mechanical properties (Figure 2). Four layers of hot-stamped sheets constituted the U-shaped component with a total thickness of 7.2 mm (the thickness of the traditional clevis-clevis is 15–20 mm). Welding and riveting were applied on the sidewall fillet areas of the four-layer sheets to achieve a rigid and stable connection. The improved structure and alternative material contributed to a strong component with significant weight reduction (above 270 kN load-bearing capacity with a nominal weight of about 2.7 kg). The traditional component bears a load of 240 kN with a nominal weight of about 7 kg [1].

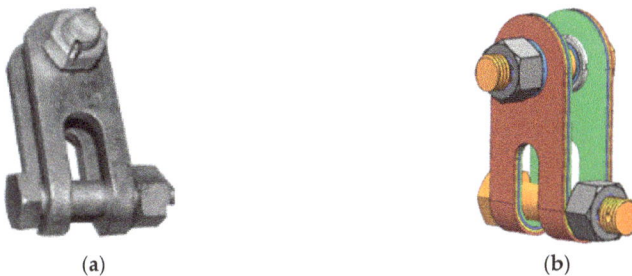

(a)   (b)

Figure 2. Structure of (**a**) the traditional clevis-clevis and (**b**) the new clevis-clevis.

The corresponding U-shaped, hot-stamping tool is shown in Figure 3. With the inclined block, the close contact between tool and sheets was obtained at the quenching stage, which resulted in a more uniform heat dissipation of the multilayered sheets.

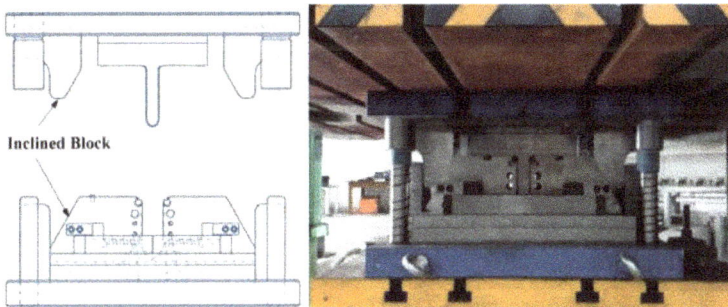

Inclined Block

Figure 3. U-shaped, multilayered-sheet, hot-stamping tool.

2.2. Numerical Analysis of the Product

The thickness of the single-layered sheet was 1.8 mm. The new type of clevis-clevis with four layers was investigated. Considering the plastic strain behaviors of the component structure, the mechanical property analysis employed an explicit dynamic method [17–19]. Furthermore, the model has the same dimensions as the new-type components, as shown in Figure 4a. This was established using ABAQUS software (version 6.13) by using element C3D8R to divide the mesh. The mechanical property analysis employed an explicit dynamic method. Zhou designed a multilayered-sheet component and established a numerical model to predict the mechanical properties of the components. However, the manufacturing process and the experimental investigation were not implemented [7]. A welded constraint was applied for all sheets at the top radius area to avoid the separation among all layers and stress concentration in the pin holes [20]. The bearing condition of the component is depicted in Figure 4b. The material of the connecting parts was SS304. It is the most common stainless steel with less electrically and thermally conductive properties than carbon steel. It has a higher corrosion resistance than regular steel with non-magnetic properties. Thus, the SS304 steel has applications in electrical connecting parts. The material of clevis-clevis was fully martensitic boron steel. The physical and mechanical parameters are shown in Table 1.

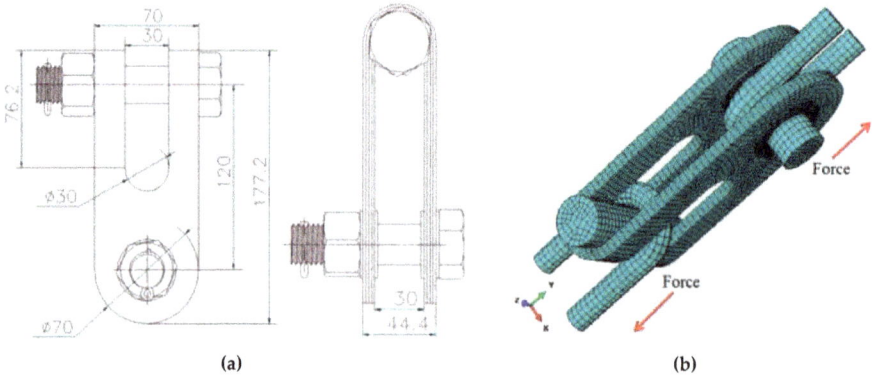

(a)   (b)

Figure 4. (**a**) The dimensions and (**b**) the finite element (FE) model of the new type of clevis–clevis. (Units: mm).

Table 1. Physical and mechanical parameters of the boron steel.

Material	Density (kg/m^3)	Modulus of Elasticity (GPa)	Poisson's Ratio	Yield Strength (MPa)	Tensile Strength (MPa)	Tangent Modulus (GPa)
Fully martensitic boron steel	7830	180	0.293	1100	1550	1.2

The repeated simulations have been implemented. In this work, the element size of 5 mm can guarantee the accuracy without a great loss of efficiency. The calculated result is shown in Figure 5. Under a working load of 80 kN, the maximum stress was 252 MPa, which is 1/3 of the yield strength, distributed in the pinhole region. The maximum stress was 657 MPa under an extreme load of 200 MPa. The stress distribution was nearly uniform in the rest of the regions under both loading conditions. The results indicated that the component can satisfy the requirement with a secure margin.

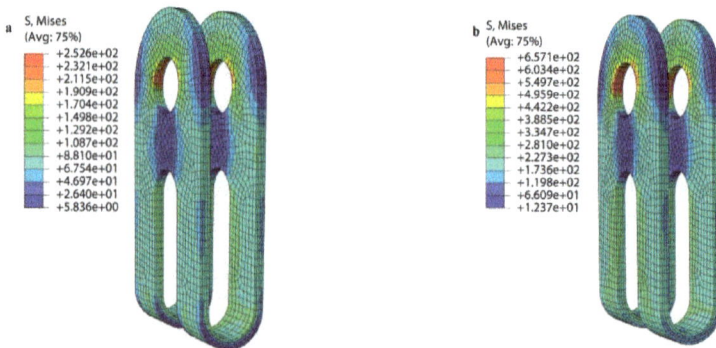

Figure 5. Equivalent stress distribution under different loads: (**a**) 80 kN and (**b**) 200 kN.

3. Multilayered-Sheet Heat Transfer Experiment

3.1. Experiment Procedure

A multilayered-sheet heat transfer experiment was conducted with a flat tool to investigate the thermal behavior of multilayered sheets during hot stamping and the quenching stages. In this test, 1.8 mm × 190 mm × 185 mm rectangular boron steel sheets were used. Table 2 shows the chemical composition of the steel. As shown in Figure 6, multilayered sheets were welded together at all corners by a resistance spot welder. Thermocouples (TC) were welded at the edge of each sheet to detect the actual temperature history of the sheet during the process. The welded joint of TC and each sheet is depicted in Figure 6. A fast-response temperature measurement and data acquisition system was designed with the National Instruments (NI) acquisition card and some intermediate processing components. The dynamic response of the system was tested, and the results showed that the system could meet the requirements of the temperature data acquisition in hot stamping and the quenching process [13]. Two flat blocks of hot work steel SKD11 with a size of 300 mm × 220 mm × 90 mm constituted the tool. The tool could be heated and maintained at the set temperature with a proportion, integral, differential controller (PID) system. The general test procedure proceeded as follows. Multilayered sheets were placed and held for 4–7 min in a furnace preheated to 1203 K. Then, the sheets were rapidly transferred to the flat die for simultaneous stamping and quenching. The dwelling time was set to 20 s. The numbers of layers were set to 1, 2, 3, and 4. Additionally, the contact pressures were set to 10, 20, 40, and 55 MPa. After hot stamping, the microscopic morphologies and tensile strength of the stamped sheets were investigated.

Table 2. Chemical composition of the 22MnB5 substrate in wt.%.

C	Mn	B	P	S	Si	Cr	Ti	Ni
0.23	1.2	0.0019	0.01	0.002	0.17	0.24	0.023	1.5

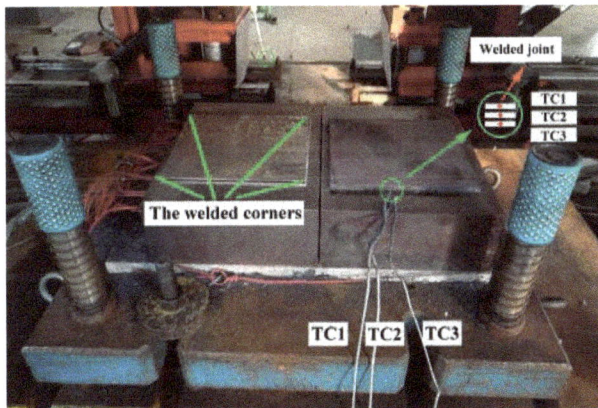

Figure 6. Schematic of the flat die experiment.

3.2. Temperature Measurement Result

The thermocouple recorded the temperature history of the sheets at the transferring, stamping, and quenching stages. Temperature curves were compared with continuous cooling transformation (CCT) of 22MnB5 steel. The temperature curves of a single-layered sheet at different contact pressures are shown in Figure 7. Although the contact pressure was as small as 10 MPa, the single-layered steel sheet cooled down rapidly at a rate of more than 30 K/s. The temperature curve declined through the single-phase area of martensitic transformation. The cooling rate increased with the contact pressure when it was smaller than 20 MPa and stabilized when the contact pressure exceeded 20 MPa.

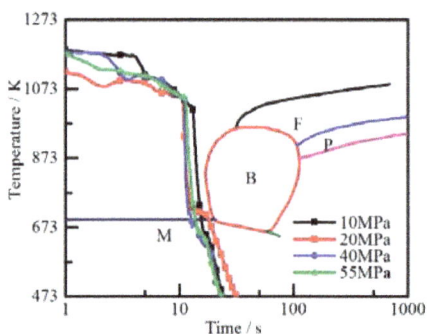

Figure 7. Temperature curves of a single-layered sheet.

The temperature curves of the double-layered sheets at different contact pressures are shown in Figure 8. The bottom and top layers are compared separately. Although the contact pressure was 10 MPa, which resulted in a relatively large contact resistance, both layers completely cooled down at a cooling rate of about 30 K/s. The layers cooled down faster when the contact pressure increased. Meanwhile, the curves indicate that the temperature behaviors of the top and bottom layers under the same pressure were basically similar.

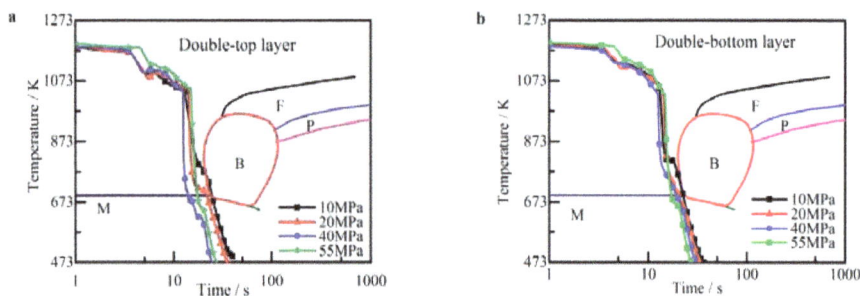

Figure 8. Temperature curves of the double-layered sheets: (**a**) top layers and (**b**) bottom layers.

For triple-layered-sheet hot stamping, thermal dissipation was slower in the middle layer than in the outer layers due to a large thermal contact resistance [15]. Whether the middle-layered sheet is completely cooled down determines the final property. The temperature curves of triple-layered sheets at different contact pressures are shown in Figure 9. The same layers from different groups are compared separately. The temperature curves of the top and bottom layers presented a larger slope than that of the middle layer. At a contact pressure of 10 MPa, the temperature curves of all three layers declined through the bainitic transformation area. However, the temperature curves of all layers declined through the edge of the bainitic transformation area at a contact pressure of 20 MPa. The temperature curve of the middle layer declined through the single-phase area of martensitic transformation at a high contact pressure of 40 MPa. Hence, bainitic transformation appeared in the final microstructure of the triple-layered sheets at a contact pressure smaller than 20 MPa. Compared with the scenario in double-layered-sheet hot stamping, contact pressure exerted a greater influence on cooling rate. Moreover, the larger pressure ensured a higher cooling rate. Hence, to obtain a full martensitic transformation, the contact pressure has to be set to more than 40 MPa.

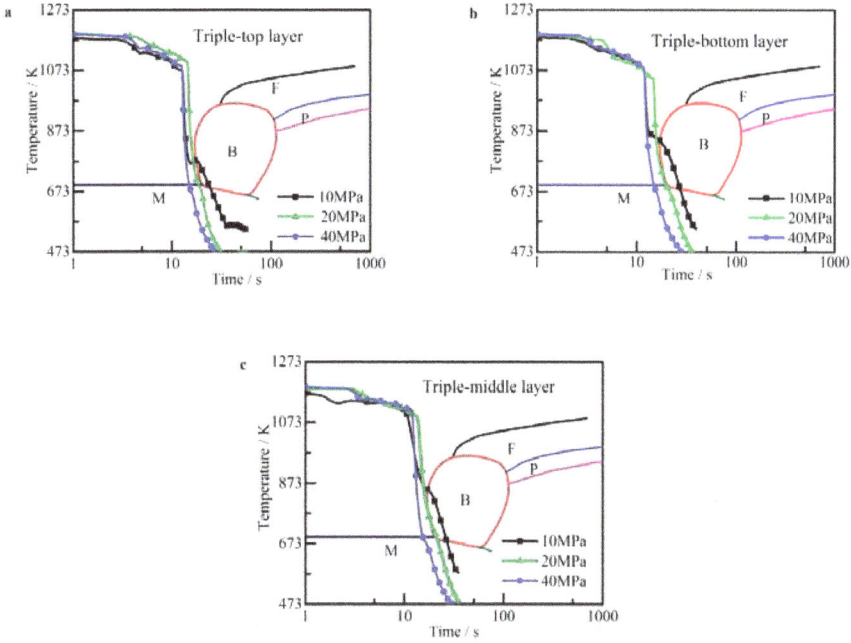

Figure 9. Temperature curves of triple-layered sheets: (**a**) top layers; (**b**) bottom layers; (**c**) middle layers.

For four-layered-sheet hot stamping, the top layer contact with the punch was defined as the first layer, and the bottom layer contact with the die was defined as the fourth layer. The rest of the sheets were named in order. The temperature curves of the third and fourth layers were obtained (corresponding to the inner- and outer-layer sheets), as shown in Figure 10. The third-layered sheet could not completely cool down even at a contact pressure of 40 MPa. Moreover, the cooling rate of the fourth layer was reduced. The difference between the curves of the third and fourth layers is apparent. Therefore, contact pressure exerted a limited influence on the cooling rate for four-layered-sheet hot stamping. Additionally, the homogenous mechanical property of the group was difficult to obtain when the number of layers was four with a total thickness of 7.2 mm.

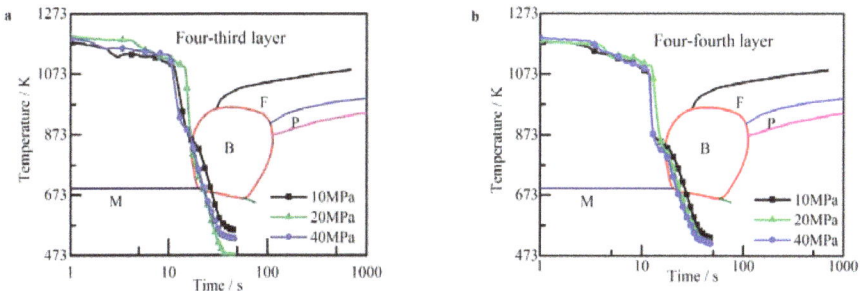

Figure 10. Temperature curves of four-layered sheets: (**a**) third layer; (**b**) fourth layer.

3.3. Microscopic Morphologies of Stamped Sheets

Metallographic specimens were cut from all sheets, mounted, ground, and polished, followed by corrosion with 3% nitric acid alcohol solution. A Quanta200 FEI scanning electron microscope was used to observe the microstructure of the specimens.

The microstructure of the top layer from the double-layered sheets at 20 MPa was set as a reference, as shown in Figure 11, in which the fine lath martensite was evenly distributed. For the triple-layered sheets at 20 MPa, as shown in Figure 12, the microstructures of the top and middle layers were a mixture of a large fraction of lath martensite, with a minor fraction of martensite–bainite being unevenly distributed. The dot martensite-bainite duplex phase accounted for a larger fraction in the middle-layered sheet than in the top-layered sheet.

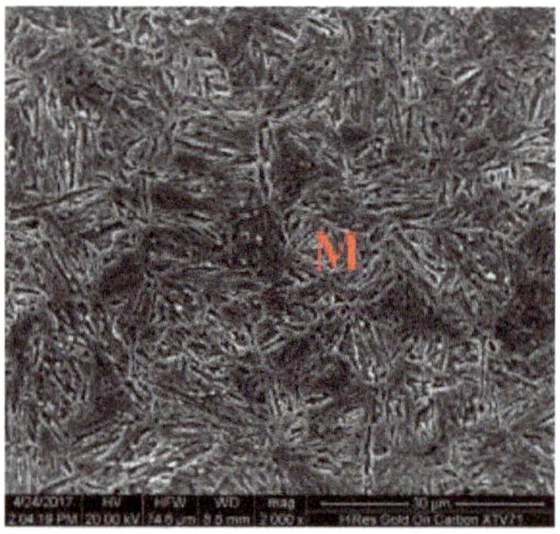

Figure 11. Microscopic morphology of the top layer from double-layered sheets at 20 MPa.

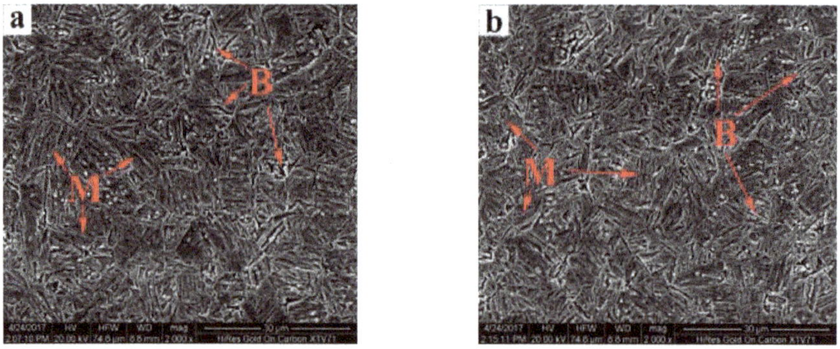

Figure 12. Microscopic morphologies of triple-layered sheets at 20 MPa. (**a**) top layer and (**b**) middle layer.

Figure 13 shows the microstructures of four-layered sheets at 20 MPa; these microstructures are quite distinct from those of the abovementioned groups. The microstructures of both layers comprised a large fraction of lath martensite, a certain fraction of bainite, and a minor fraction of ferrite. The difference between each layer was negligible. This result is the consequence of the low cooling rates of all four layers.

Comparison of the microscopic morphologies and temperature curves revealed the following facts. For multi-layered-sheet hot stamping, when the number of layers is less than three, contact pressure affects the cooling rate, which determines the microstructure of the stamped sheets. When the number of layers is four, the influence of contact pressure on the cooling rate is modest, such that bainitic transformation is inevitable under experimental conditions.

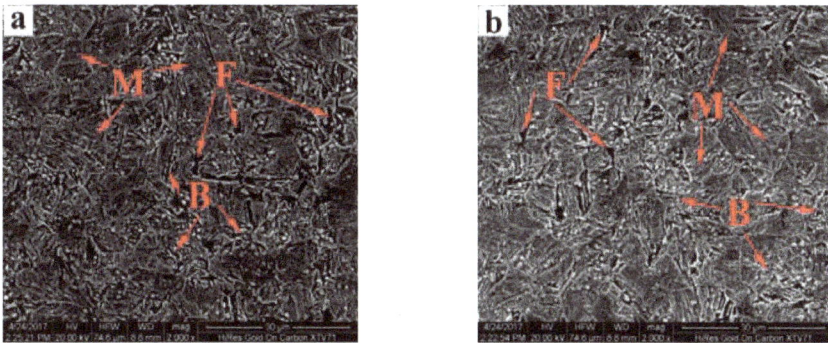

Figure 13. Microscopic morphologies of four-layered sheets at 20 MPa, (**a**) third layer and (**b**) fourth layer.

3.4. Tensile Tests

To further verify the mechanical properties of the stamped sheets, tensile tests were conducted. Specimens for tensile testing were machined according to DIN 50114. The dimension of specimens for tensile tests at room temperature is shown in Figure 14. The tensile tests were carried out on a SHIMADZU AG- 100 kN machine with a deformation rate of 2 mm/min. The stress and deformation values were collected during the tests. The experimental results are shown in stress–strain curve.

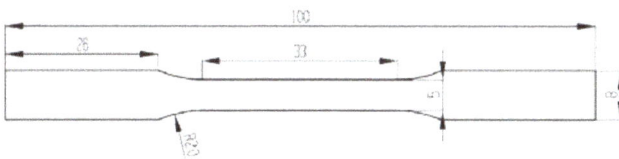

Figure 14. Dimension of specimens for tensile test at room temperature. (Unit: mm).

The engineering stress versus strain curves of the single-layered sheet at different contact pressures are shown in Figure 15. Even at a small contact pressure of 10 MPa, the stamped sheet with a fully martensitic microstructure exhibited a yield strength of about 1200 MPa and a tensile strength of above 1500 MPa. Hence, the single-layered sheet was completely cooled down during the stamping process.

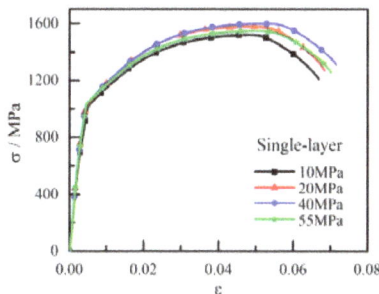

Figure 15. Stress-strain curves of the single-layered sheet.

The engineering stress versus strain curves of the double-layered sheets are shown in Figure 16. The strength of the double-layered sheets had the same strength grade as that of the single-layered sheet. Therefore, the cooling rate was high enough to completely cool down the 3.6 mm stacked sheets. The difference between the top- and bottom-layer sheets was negligible.

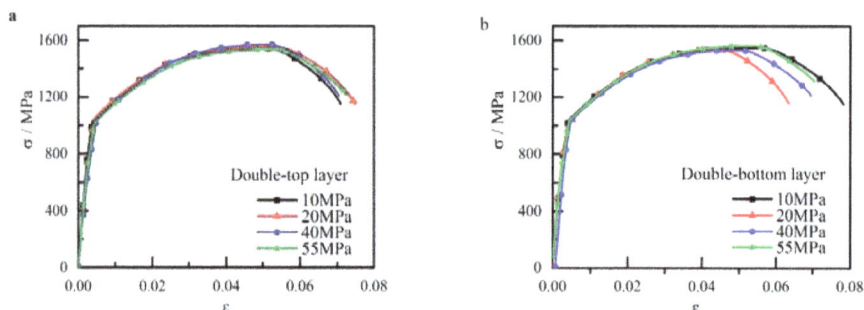

Figure 16. Stress-strain curves of the double-layered sheets. (**a**) top layer and (**b**) bottom layer.

The engineering stress versus strain curves of the triple-layered sheets are shown in Figure 17. The tensile strength of all layers was about 1400 MPa at a contact pressure of 10 MPa, and the strength increased as the contact pressure increased. At a contact pressure of 40 MPa, the tensile strength reached 1450 MPa. Moreover, the tensile strength of each layer from the same group had a compatible strength grade of 1400 MPa. Hence, the slight difference of the dot martensite-bainite duplex phase barely affected the final property. In contrast to the microstructure of the double-layered sheets, the appearance of the duplex phase resulted in a slight strength reduction. Thus, contact pressure was still the determining factor in the triple-layered-sheet hot-stamping process.

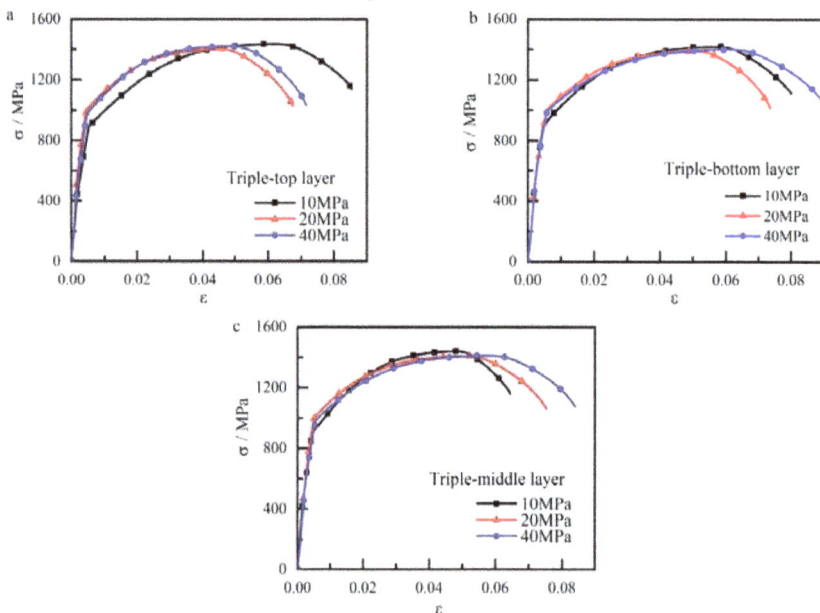

Figure 17. Stress–strain curves of triple-layered sheets. (**a**) Top layer, (**b**) bottom layer, and (**c**) middle layer.

The engineering stress versus strain curves of the four-layered sheets are shown in Figure 18. Every layer from top to bottom was named in order as mentioned above (the layer contact with the punch was the first layer, and the layer contact with the die was the fourth layer). The tensile strength of each layer was less than 1150 MPa even at a pressure of 40 MPa. A certain amount of bainitic transformation significantly influenced the strength reduction. Additionally, the outer layers directly in contact with the tool obtained higher strength than the middle layers; this finding is attributed to the relatively faster cooling rate with a smaller fraction of bainite in the outer layers.

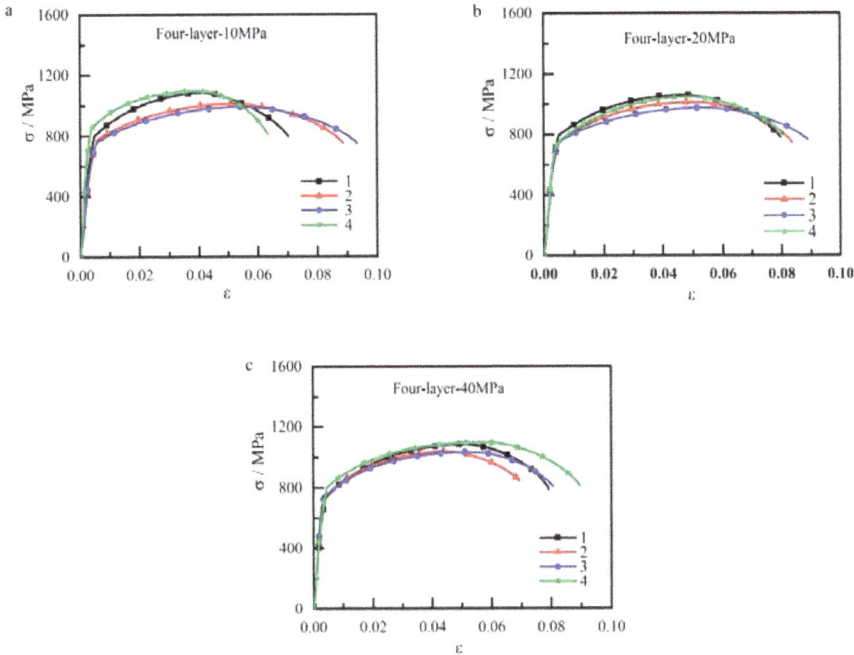

Figure 18. Stress-strain curves of the four-layered sheets: (**a**) 10 MPa, (**b**) 20 MPa, and (**c**) 40 MPa.

In conclusion, during multilayered-sheet hot stamping, single- and double-layered sheets with a total thickness of less than 3.6 mm could be completely cooled down even at a small contact pressure of 10 MPa. In triple-layered sheets with a total thickness of 5.4 mm, the martensite-bainite duplex phase was unevenly distributed in the stamped sheets at a contact pressure of 20 MPa, which resulted in a slight strength reduction and inhomogeneous properties. To obtain a critical cooling rate of 30 K/s, the contact pressure must be set to more than 40 MPa. While the number of layers is four with a total thickness of 7.2 mm, all sheets could not be cooled down. A certain fraction of bainitic transformation appeared in all sheets at a large contact pressure of 40 MPa, which resulted in a large strength reduction. Hence, to obtain the fully martensitic microstructure, the optimizing parameters for the U-shaped, multilayered-sheet, hot-stamping process are double-layered sheets at a contact pressure of above 20 MPa.

4. New Product and Its Performance Tests

The U-shaped tool can manufacture two parts of a component in one hot-stamping process. The parameters of the U-shaped stamping process were concluded from the above experiment. In this test, prepared sheets with a size of 3.6 mm × 310 mm × 70 mm were used. The double-layered sheets were welded together at all corners by a resistance spot welder. The chemical composition of the steel sheet is shown in Table 2. The double-layered sheets were placed and held for 4–7 min in a furnace preheated to 1203 K. Then, the sheets were rapidly transferred to the U-shaped tool for simultaneous stamping and quenching. The contact pressure was set to 50 MPa. The dwelling time was set to 20 s. Then, the two formed parts were stacked (Figure 19a) and spot-welded at the edges. Then, the pinholes were connected with the bolt, as shown in Figure 19b.

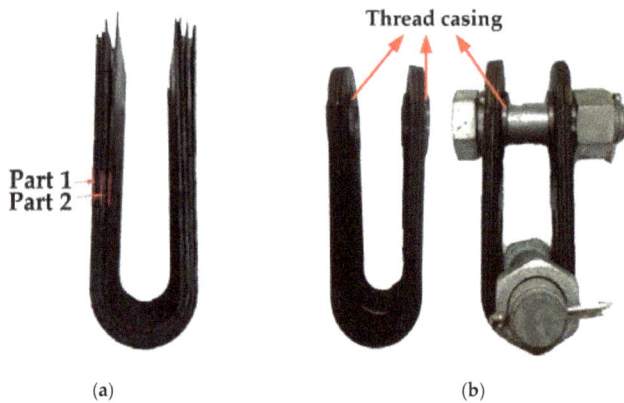

Figure 19. (a) The formed parts and (b) new type of clevis-clevis.

For the new component to meet the national standards of electrical power fittings, the component has to bear 1.2 times of the nominal load without damage, namely 240 kN. The arranged accumulated stress was close to the real work condition. The test procedure was divided into three steps. First, the fitting at both clamps was fixed, loading the fitting to a tensile stress of 200 kN. Then, it was held for 60 s and unloaded. Finally, the loading and holding process was repeated, and it was then loaded to 240 kN and held for 60 s. The fitting is qualified if no fracture appeared. Testing with a dedicated device in an electrical power fitting company in the south of China indicated that when the tension reached 240 kN and even 280 kN, this new type of clevis-clevis exhibited no fracture failure. The experimental diagram of product performance tests is shown in Figure 20.

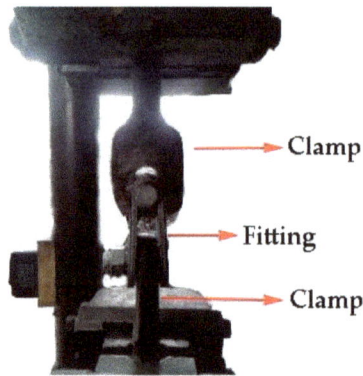

Figure 20. The experimental diagram of product performance tests.

5. Conclusions

(1) A multilayered-sheet, hot-stamping process was proposed. A model was designed and used to produce an electric-power-fitting product clevis-clevis component by multilayered-sheet hot stamping.

(2) The number of layers and contact pressure were the key process parameters in multilayered-sheet hot stamping; they determined the final microstructure and mechanical properties. When the number of sheets was two and the total thickness was 3.6 mm, each sheet could obtain a fully martensitic microstructure at a relatively low contact pressure of 10 MPa. When the number of layers was three and the total thickness was 5.4 mm, a relatively high contact pressure of more than 40 MPa was needed

to ensure an adequate cooling rate. When the number of layers is four, the inner layer could not cool down even at a contact pressure of more than 40 MPa.

(3) U-shaped, double-layered-sheet hot stamping was implemented to produce a typical electrical connection fitting clevis-clevis. The bearing capacity of the four-layered clevis-clevis was tested. The new clevis-clevis component exhibited a large load capacity of 280 kN, which satisfies the requirement with a secure margin. Compared with that of the traditional component, the weight of the new component was reduced by 60%.

(4) Considering actual application status, corrosion protection of the clevis-clevis component is essential [21], and hot galvanization on the surfaces of clevis-clevis to prevent corrosion and reliability for creep deformations of the component will be investigated in future research.

Author Contributions: Conceptualization, B.Z.; Methodology, Y.W. and Y.Z.; Software, Y.J.; Validation, K.W.; Investigation, Z.Z.; Resources, Y.W. and Y.Z.

Funding: This research work was funded by the National Natural Science Foundation of China (grant No. 5140517 and U1564203).

Acknowledgments: The authors would like to acknowledge the Analytic and Testing Center of the State Key Lab of Materials Processing and Die and Mould Technology for their assistance in the tensile tests.

Conflicts of Interest: The authors declare no conflict of interest.

References

1. Meah, K.; Ula, S. Comparative Evaluation of HVDC and HVAC Transmission Systems. In Proceedings of the Power Engineering Society General Meeting, Tampa, FL, USA, 24–28 June 2007.
2. Liu, S.; Fan, Z.; Gao, M. Analysis of the reasons behind the fracture of the 220kV pipe busbar horizontal line clamp. In Proceedings of the 2016 International Conference on Applied Engineering, Materials and Mechanics (ICAEMM 2016), Weihai, China, 15–17 January 2016.
3. Karbasian, H.; Tekkaya, A.E. A review on hot stamping. *J. Mater. Process. Tech.* **2010**, *210*, 2103–2118. [CrossRef]
4. Maider, M.; Garikoitz, A.; Anton, G.; Carlos, A. Effect of the martensitic transformation on the stamping force and cycle time of hot stamping parts. *Metals* **2018**, *8*, 385.
5. Mulidrán, P.; Šiser, M.; Ján, S.; Spišák, E.; Sleziak, T. Numerical prediction of forming car body parts with emphasis on springback. *Metals* **2018**, *8*, 435. [CrossRef]
6. Hall, J.N.; Fekete, J.R. *Steels for Auto Bodies: A General Overview*, 1st ed.; Cambridge Woodhead Publishing Press: Cambridge, UK, 2017; pp. 19–45.
7. Zhou, M.L.; Wang, L.; Wang, Z.J.; Wang, Y.L.; Zhang, Y.S. Analysis of the design and mechanical performance about hot stamping power fitting products. *Adv. Mater. Res.* **2014**, *1063*, 272–275. [CrossRef]
8. Chang, Y.; Meng, Z.H.; Ying, L.; Xiao-Dong, L.I.; Ning, M.A.; Ping, H.U. Influence of hot press forming techniques on properties of vehicle high strength steels. *J. Iron Steel Res.* **2011**, *18*, 59–63. [CrossRef]
9. Yoon, S.H.; Kyu, J.C.; Gil, K.C. Effect on blank holding force on blank deformation at direct and indirect hot deep drawings of boron steel sheets. *Metals* **2018**, *8*, 574.
10. Seung, H.L.; Jaewoong, P.; Kiyoung, P.; Dong, K.K.; Hyunwoo, L.; Daeho, Y.; Hongrae, P.; Jaeseung, K. A study on the cooling performance of newly developed slice die in the hot press forming process. *Metals* **2018**, *8*, 947.
11. Hay, B.A.; Bourouga, B.; Dessain, C. Thermal contact resistance estimation at the blank/tool interface: Experimental approach to simulate the blank cooling during the hot stamping process. *Int. J. Mater. Form.* **2010**, *3*, 147–163.
12. Abdulhay, B.; Bourouga, B.; Dessain, C.; Brun, G.; Wilsius, J. Development of estimation procedure of contact heat transfer coefficient at the part-tool interface in hot stamping process. *Heat Transfer Eng.* **2011**, *32*, 497–505. [CrossRef]
13. Wang, C.; Zhang, Y.S.; Tian, X.W.; Zhu, B.; Li, J. Thermal contact conductance estimation and experimental validation in hot stamping process. *Sci. China Technol. Sci.* **2012**, *55*, 1852–1857. [CrossRef]
14. Zhu, B.; Zhang, Y.S.; Li, J.; Wang, H.; Ye, Z. Simulation research of hot stamping and phase transition of automotive high strength steel. *Material. Res. Innov.* **2015**, *15*, s426–s430. [CrossRef]

15. Wang, Z.J.; Xu, Y.; Zhou, M.L.; Wang, Y.L.; Zhang, Y.S. Valuation method for effects of hot stamping parameters on tailored properties. *Adv. Mater. Res.* **2014**, *1063*, 202–206. [CrossRef]
16. Gui, Z.X.; Zhang, Y.S.; Wang, Z.J. Simulation of lightweight B-pillar during hot stamping process with thermo-mechanical-metallurgical model. *Stamping Techn.* **2012**, *37*, 193–197.
17. Lee, M.G.; Kim, S.J.; Han, H.N.; Jeong, W.C. Application of hot press forming process to manufacture an automotive part and its finite element analysis considering phase transformation plasticity. *Int. J. Mech. Sci.* **2009**, *51*, 888–898. [CrossRef]
18. Bok, H.H.; Lee, M.G.; Pavlina, E.J.; Barlat, F.; Kim, H.D. Comparative study of the prediction of microstructure and mechanical properties for a hot-stamped B-pillar reinforcing part. *Int. J. Mech. Sci.* **2011**, *53*, 744–752. [CrossRef]
19. Tang, B.; Wang, Q.; Wei, Z.; Meng, X.; Yuan, Z. FE simulation models for hot stamping an automobile component with tailor-welded high-strength steels. *J. Mater. Eng. Perform.* **2016**, *25*, 1709–1721. [CrossRef]
20. Yanzhong, J.U.; Jiang, F.; Dehong, W.; Lili, M.; Shuzhen, Z. Ultimate bearing capacity analysis of safety pin based on nonlinear contact. *Water Resour. Power* **2012**, *7*, 173–175.
21. DžUpon, M.; Falat, L.; Slota, J.; Hvizdoš, P. Failure analysis of overhead power line yoke connector. *Engineering Fai. Anal.* **2013**, *33*, 66–74. [CrossRef]

![metals logo] *metals*

MDPI

Article

Modification of Microstructure and Texture in Highly Non-Flammable Mg-Al-Zn-Y-Ca Alloy Sheets by Controlled Thermomechanical Processes

Sangbong Yi [1,*], José Victoria-Hernández [1], Young Min Kim [2], Dietmar Letzig [1] and Bong Sun You [2]

[1] Institute of Materials Research, Helmholtz-Zentrum Geesthacht, Max-Planck-Str. 1, 21502 Geesthacht, Germany; jose.victoria-hernandez@hzg.de (J.V.-H.); dietmar.letzig@hzg.de (D.L.)
[2] Implementation Research Division, Korea Institute of Materials Science, 797, Changwon-daero, Seongsan-gu, Changwon 51508, Korea; ymkim@kims.re.kr (Y.M.K.); bsyou@kims.re.kr (B.S.Y.)
* Correspondence: sangbong.yi@hzg.de; Tel.: +49-(0)4152-871911

Received: 28 December 2018; Accepted: 28 January 2019; Published: 2 February 2019

Abstract: The influence of rolling temperature and pass reduction degree on microstructure and texture evolution was investigated using an AZXW3100 alloy, Mg-3Al-1Zn-0.5Ca-0.5Y, in wt.%. The change in the rolling schedule had a significant influence on the resulting texture and microstructure from the rolling and subsequent annealing. A relatively strong basal-type texture with a basal pole split into the rolling direction was formed by rolling at 450 °C with a decreasing scheme of the pass reduction degrees with a rolling step, while the tilted basal poles in the transverse direction were developed by using an increasing scheme of the pass reduction degrees. Rolling at 500 °C results in a further distinct texture type with a far more largely tilted basal pole into the rolling direction. The directional anisotropy of the mechanical properties in the annealed sheets was caused by the texture and microstructural features, which were in turn influenced by the rolling condition. The Erichsen index of the sheets varied in accordance to the texture sharpness, i.e., the weaker the texture the higher the formability. The sheet with a tetrarchy distribution of the basal poles into the transverse and rolling directions shows an excellent formability with an average Erichsen index of 8.1.

Keywords: magnesium alloys; texture; formability; rolling

1. Introduction

In the last decades magnesium (Mg) alloys have been widely investigated due to their positive characteristics for a lightweight structure, e.g., their low density, high machinability and excellent damping capacity. However, the poor formability of Mg sheets, especially at room temperature, is one of the main drawbacks that retards the industrial application of semi-finished products. Conventional wrought Mg alloys, e.g., those based on the Mg-Al-Zn system such as the AZ31 alloy, have a tendency to develop strong basal-type textures during the sheet rolling process. The basal-type texture, in which most grains have their c-axes in the sheet normal direction (ND), causes a limited sheet formability from the restricted activities of the <a> dislocations slip, especially for the strain accommodation along the ND. The pyramidal <c+a> slip with a Burgers vector of <11$\overline{2}$3> can accommodate the strain along the ND of such strongly textured material. Nevertheless, the <c+a> slip can be activated only at high temperature due to the high critical resolved shear stress (CRSS) at room temperature. To improve the formability of the Mg sheets, it is essential to provide a way of weakening the texture. In case of a weakened basal-type texture or non-basal texture components, it is expected that <a> dislocations with a relatively low CRSS contribute to accommodating the deformation along the ND. It was reported that texture weakening can be achieved by alloying Mg with yttrium (Y) and rare earth (RE) elements such as cerium (Ce) or neodymium (Nd) [1–5]. Indeed, increased ductility, formability and strength have been observed in the sheet that has a weaker texture [6]. Even though the alloying addition of RE elements is an effective method of weakening textures and improving sheet formability at low temperature, the Mg alloy sheets with a reduced alloying amount or without RE elements, which are strategically important raw materials, are very beneficial from ecological and economic viewpoints.

The addition of Ca into Mg alloys is known to improve high temperature mechanical properties and ignition-proof behavior. Especially, the simultaneous addition of Ca with Y results in an excellent ignition-proof behavior and improved corrosion resistance. The Mg alloys with significantly improved non-flammability have attracted much attention for research activities and industrial applications [7–10]. Recent studies have shown that Ca addition also alters the strong basal-type texture to a more randomly distributed orientation after thermomechanical treatments, such as extrusion [11] or rolling [12,13], and, consequently improves room temperature formability. The Mg alloys containing Ca can be further strengthened by an optimized aging hardening scheme [14,15]. It is to be mentioned that the Ca added alloys investigated for wrought processes mostly handle with Al-free Mg alloys, e.g., Mg-Zn-Ca and Mg-Mn-Ca systems. Regarding further properties such as corrosion resistance and non-flammability, it is important to provide a method of texture weakening with a fine and homogeneous grain structure simultaneously in various alloy systems, especially in Mg alloys with Al addition. Moreover, studies on microstructure evolution during sheet processing and the mechanical behavior of the newly developed non-flammable alloy containing Ca and Y is limited.

In the present study, the microstructure and formability of the non-flammable AZXW3100 sheets, which are a modified AZ31 alloy by Ca and Y addition produced by different hot rolling schedules, were examined. A focus of the present study is to analyze the relationship between the thermomechanical process, microstructure evolution, and resulting properties.

2. Materials and Methods

The ingots of AZXW3100 alloy, with a nominal composition of Mg-3Al-1Zn-0.5Ca-0.5Y in wt.%, were cast into a steel mold under an Ar and SF6 atmosphere. The corresponding alloy compositions are well recognized for their excellent ignition resistance and high strength after extrusion [7,9], while their microstructure evolution during rolling and the resulting sheet properties have not been systematically studied. The cast ingot slabs with a thickness of 10 mm were machined and homogenized at 450 °C for 20 h under a continuous Ar flow. The homogenized slabs were rolled to a final gauge of 1 mm using different rolling conditions that are shown in Table 1. Two different schemes of deformation degree per pass were applied at the rolling temperatures of 450 °C and 500 °C; the deformation degrees per pass increased with the rolling step (φ = 0.1 to 0.3) or decreased with the rolling step (φ = 0.3 to 0.1).

Table 1. Different rolling conditions used in the present study.

Rolling Condition	Rolling Temperature (°C)	Deformation Degree per Pass (φ) (Total 11 Rolling Steps)	Remarks
450-inc	450	0.1 → 0.3 (increasing with rolling step)	
450-dec	450	0.3 → 0.1 (decreasing with rolling step)	$\varphi = -\ln\left(t_{(n+1)} / t_n\right)$ where,
500-inc	500	0.1 → 0.3 (increasing with rolling step)	t_n = sheet thickness at the nth rolling step.
500-dec	500	0.3 → 0.1 (decreasing with rolling step)	

The rolled sheets were annealed at 400 °C for different lengths of time, from 300 to 3600 s. The microstructure evolution during the rolling and recrystallization annealing were investigated with focus given to the influence of the different rolling schemes. Optical microstructures of the rolled and annealed sheets were observed by using standard metallographic sample preparation techniques and an etchant based on picric acid [16]. Global texture measurements on the rolled and recrystallized sheets were conducted using X-ray diffraction (Cu Kα, 40 kV and 40 mA, Panalytical, Almelo, The Netherland). The orientation distribution function and complete pole figures were calculated by using a MATLAB toolbox MTEX [17] from six measured pole figures. Specimens for electron backscatter diffraction (EBSD) analysis were prepared by electropolishing in a Struers AC2 solution at −20 °C and 30 V. The EBSD measurements were conducted on a field emission gun scanning electron microscope (working at 15 kV, Zeiss Ultra 55, Carl Zeiss AG, Oberkochen, Germany) equipped with a Hikari detector (AMETEK Inc., Mahwah, NJ, USA) and an EDAX/TSL EBSD system.

The tensile samples were prepared from the sheets after recrystallization annealing in 3 different sheet directions: rolling direction (RD), transverse direction (TD) and 45° to RD, according to the DIN 50125 H12.5 × 50. The quasi-static tensile tests were conducted at room temperature with an initial strain rate of 10^{-3}/s. The stretch formability of the sheets was examined by Erichsen tests of the as-rolled and annealed sheets at room temperature. The tests were performed using a punch with a diameter of 20 mm at a punch speed of 5 mm/min and a blank hold force of 10 kN, according to DIN 50101. The Erichsen index (IE) was determined by the punch stroke corresponding to the max load. The results from the tensile tests and the Erichsen tests are given as the average values from 3 samples, at least, for each condition of the examined sheet.

3. Results and Discussion

The optical microstructures, which are taken from longitudinal sections, and the recalculated (0001) and {10-10} pole figures of the AZXW3100 alloy sheets rolled with different conditions are shown in Figure 1, Figure 2, and Figure 5. The microstructural evolution during the recrystallization annealing at 400 °C of each sheet is presented. The numbers given on the micrographs indicate the average grain sizes measured by the linear intercept method.

The sheet rolled with the 450-inc rolling scheme exhibits a strongly deformed structure with a large number of twin and deformation bands, which are homogeneously distributed in the whole sheet (Figure 1a). The grains are elongated in the RD. The average grain size of the rolled sheet is about 11 μm, which is determined from the matrix grains without counting the twin boundaries. The annealed sheet at 400 °C for 300 s shows a fully recrystallized microstructure with equi-axed grains and an average size of 6 μm. The recrystallized grain structure is very stable and no significant grain growth occurs during the subsequent annealing, such that the grain size of the annealed sheet for 3600 s at 400 °C is 8 μm. Many of the fragmented secondary phases are aligned along the RD. They are observed in the as-rolled as well as the recrystallized sheets.

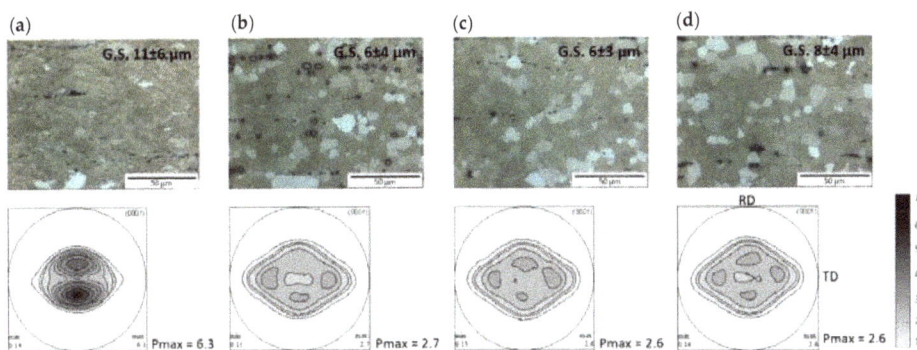

Figure 1. The optical micrographs and the recalculated (0002) pole figures of the sheets rolled by the 450-inc rolling scheme, i.e., the increasing deformation degrees with rolling step, (**a**) at the as-rolled condition and after annealing for (**b**) 300 s, (**c**) 600 s and (**d**) 3600 s at 400 °C.

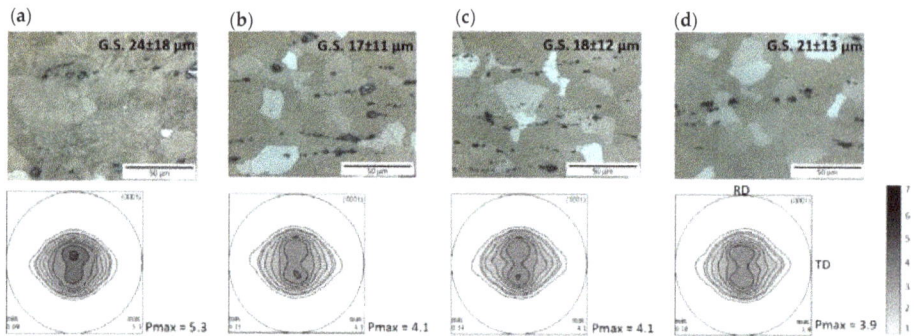

Figure 2. The optical micrographs and the recalculated (0002) pole figures of the sheets rolled by the 450-dec rolling scheme, i.e., the decreasing deformation degrees with rolling step, (**a**) at the as-rolled condition and after annealing for (**b**) 300 s, (**c**) 600 s and (**d**) 3600 s at 400 °C.

A relatively strong texture with a max pole density of the (0001) pole figure, Pmax = 6.3 multiple random distribution (m.r.d.), is obtained in the sheet rolled with the 450-inc scheme. The as-rolled sheet shows the basal pole distribution splitting into the RD. The formation of the texture component is understood as a result from a high activity <c+a> slip and secondary twinning during deformation [18]. The recrystallization annealing leads to a texture weakening such that the max pole density of the annealed sheet is Pmax = 2.6 m.r.d. Texture weakening accompanies the development of the basal pole split into the TD, while the basal poles in the RD become significantly weaker. This tendency of texture development, i.e., recrystallization texture with an additional basal pole split into the TD, has been mostly reported in studies of various RE or Ca containing Mg alloys sheets. It is important to note that the formation of weak textures with a TD basal pole split has been reported mostly in Al-free Mg alloys, such as Mg-Zn-RE and Mg-Zn-Ca systems. The present results show that such texture components are also obtained in Al-containing alloys by controlling the thermomechanical treatment conditions. This type of sheet texture stands out among others for its more homogeneous orientation distribution, which is usually more desirable for sheet metal forming processes than a texture with a predominant distribution of basal poles along one sheet direction.

Changing the rolling scheme produced a significantly different microstructure and texture in the final sheet. When decreasing the deformation degree with a rolling step, i.e., 450-dec rolling scheme with rolling degrees of $\varphi = 0.3$ at the beginning and $\varphi = 0.1$ at the final rolling steps, the rolled sheet showed a coarse grain structure with an average grain size of 24 μm. A fully recrystallized microstructure was shown after 300 s annealing at 400 °C, and the average grain size slightly varied from 18 μm to 21 μm during the recrystallization annealing up to 3600 s. The grain structure is relatively inhomogeneous, as shown by the relatively large values of the grain size deviation.

The as-rolled sheet shows the texture with a spread of the basal poles, up to about 20° tilting into the RD. That is, most grains have their basal planes aligning parallel to the sheet normal plane. The recrystallization annealing accompanied a slight weakening of the basal-type texture, from Pmax = 5.1 m.r.d to Pmax = 4.1 m.r.d. at the as-rolled and recrystallized conditions, respectively. During the recrystallization annealing up to 3600 s at 400 °C the texture with the basal pole spread into the RD was maintained without a qualitative change. The sheet rolled by the 450-dec scheme shows a relatively stronger recrystallization texture, Pmax = 3.9 m.r.d. after 3600 s annealing, than the sheet rolled with the 450-inc scheme. Due to the low deformation degree at the final rolling steps the coarse grains formed during the intermediated annealing were kept without fragmentation by deformation, e.g., twinning and deformation bands. The grain structure and texture development during the post annealing of the 450-dec sheet seems to be related to the extended recovery-controlled process, such that no significant texture change resulted from the annealing. In contrast, the sheet rolled with a large deformation degree at the final rolling steps, e.g., the sheet rolled with the 450-inc scheme, has a high deformation energy at the as-rolled condition. Thus, the recrystallization, including the nucleation, which is triggered mostly at highly deformed areas, like deformation bands, shear bands and grain boundaries, occurs during the annealing simultaneously with the extended recovery. This recrystallization process results in qualitative texture changes that accompany the weakening of deformation texture components.

The EBSD measurements clearly indicate the differences in the microstructure and deformation energy of the as-rolled sheets (Figure 3). The inverse pole figure (IPF) map and the kernel average misorientation (KAM) map of the EBSD measurement show the grain orientations and degree of deformation at measuring points. The KAM map is well acknowledged to represent the stored energy and geometrically necessary dislocations (GND) remaining in a deformed material. The rolled sheet by the 450-inc scheme, Figure 3a, shows a large number of twins, which correspond to the $\{10-12\}-\{10-11\}$ secondary twins determined by its orientation relationship to the matrix [19]. Moreover, the KAM map demonstrates that a relatively high deformation energy is distributed within the whole volume at the grain interior and along the grain boundaries, where a relatively higher deformed zone is observed along the lines inclined about 30° from the RD. The band-shaped structure is observed in the as-rolled sheet, marked with grey lines in Figure 3a. The band structure, at which the deformation energy is more concentrated, mostly develops along the twins and seems to be a microstructural feature prior to the shear band. The sheet rolled by the 450-dec scheme shows a deformed microstructure with twins. However, the degree of deformation is much lower than that of the 450-inc sheet. The KAM map also indicates that the deformation degree is more or less concentrated along the grain and twin boundaries. The elements distribution maps at the same area to the EBSD measurement of the 450-dec sheet are shown in Figure 4. It is obvious that the secondary phases observed in the examined alloy, which are shown as black dots in the EBSD maps, are mostly the high Al-Ca and Al-Y phases, e.g., $(Al, Mg)_2Ca$ and Al_4MgY. A relatively high concentration of Mn, 2~5 wt.% from the EDX analysis, was found at the $Al4Mg_4Y$ phase.

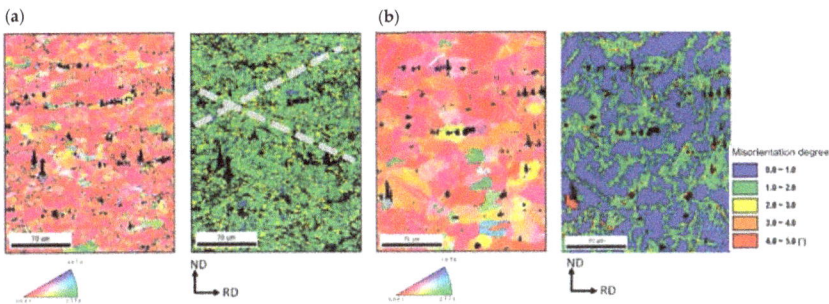

Figure 3. EBSD inverse pole figure maps and kernel average misorientation maps of the as-rolled sheets by (**a**) 450-inc and (**b**) 450-dec rolling schemes.

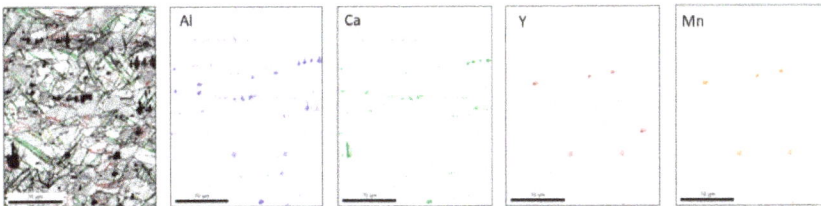

Figure 4. Grain boundary map and alloying elements distribution maps the as-rolled sheet by 450-dec rolling scheme, at the same measuring area with the Figure 3b. Different colors of the boundaries, in the grain boundary map, correspond to the twin types: red to $\{10-12\}$ extension, yellow to $\{10-11\}$ contraction and green to $\{10-12\}-\{10-11\}$ secondary twins.

The influence of the rolling scheme with regards to the step reduction degree on the microstructure and texture evolution is negligible, when the sheet rolling is conducted at a higher temperature. The sheets rolled at 500 °C have a distinct texture type, where the basal pole split into the RD with a tilting angle of 20~30° from the ND is formed independently from the step reduction degree (Figure 5a,c). After annealing for 600 s at 400 °C, the texture weakens to Pmax = 3.6 and 3.4 in the 500-inc and 500-dec sheets, respectively. This type of texture with the basal pole split into the RD and a relatively weaker intensity compared to a commercial Mg alloy is generally observed in the Mg-REE binary alloy sheets [20]. The grain sizes of the annealed sheets, 9 μm and 7 μm, are comparable to those of the 450-inc sheet. Besides the intermetallic particles, many stringer structures of oxide inclusions, identified by means of EDX analysis, are found along the RD in the optical micrographs, and also in the sheets rolled at 450 °C. These microstructural features originate from the casting defects, e.g., micro-voids and oxide inclusions, which are elongated during the rolling procedures.

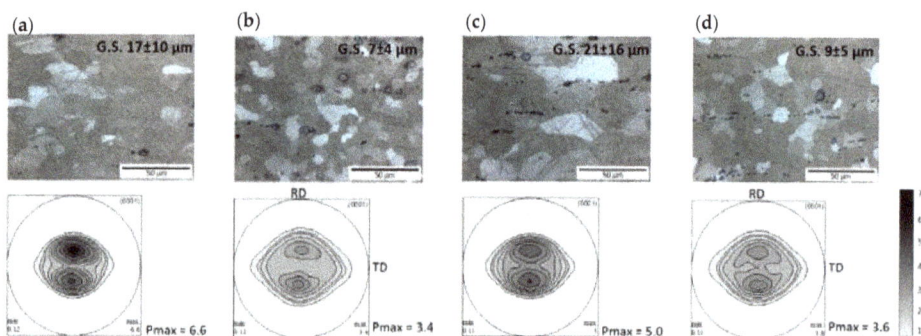

Figure 5. The optical micrographs and the recalculated (0002) pole figures of the sheets rolled by the 500-inc and 500-dec schemes, (**a**), (**c**) as-rolled condition and (**b**), (**d**) after annealing for 600 s at 400 °C, respectively.

Figure 6 displays the stress strain curves during the uniaxial tensile tests of the annealed 450-inc and 450-dec sheets in three sheet planar directions, at RD, TD, and 45° from the RD. The mechanical properties of the sheets are listed in Table 2. The higher strength and ductility of the 450-inc sheet are associated with a finer grain structure and weaker texture than those of the 450-dec sheet. Due to this weak texture, i.e., the more randomly distributed orientations, the dislocation slip can be more easily activated without a geometrical restriction, i.e., a higher Schmid's factor for the basal <a> slip than a more strongly textured material, such that higher ductility is achieved based on a homogeneous deformation. The 450-inc sheet has a high yield strength ($\sigma_{0.2}$ > 150 MPa) in different sheet planar directions. Moreover, other alloys sheets that have a similar texture, e.g., the ZE10 sheet in [21], have a much lower strength, especially in the TD. The anisotropic mechanical properties of the sheets can also be understood from the texture. The higher stress values in the RD than the TD are associated with the broader angular tilt of the basal poles towards the TD than the RD. Due to its geometrical advantages, the basal <a> slip is more favored during the tensile loading in the TD. The directional anisotropy in the yield strength is larger in the 450-inc sheet, which is attributed to the basal pole split into the TD. The 450-dec sheet shows a basal pole spread into the TD and, accordingly, the smaller directional anisotropy of the yield strength is observed. Indeed, it is expected that the anisotropy in ductility is insignificant, even with a slightly higher ductility in the TD [21]. In general, the sheet planar direction with lower stress corresponds to higher uniform strain and fracture strain. The premature fracture during the tension in the TD of the present study was caused by the oxide inclusions aligned along the RD. The stringers of the oxide inclusions act as a stress concentration site, especially during the loading along the TD, such that the fracture occurs at a low strain level, as also shown in a previous study [22].

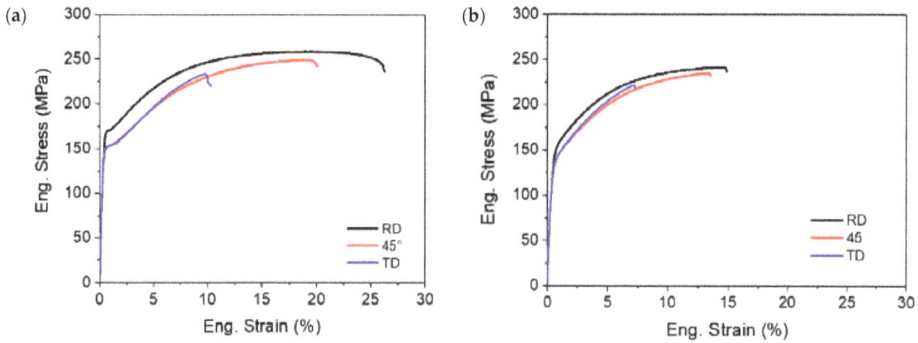

Figure 6. Stress strain curves of the annealed sheets for 600 s at 400 °C after rolling by (**a**) 450-inc and (**b**) 450-dec schemes.

Table 2. Mechanical properties of the annealed sheets from tensile tests.

Tensile Properties	Annealing for 600 s at 400 °C					
	450-inc			450-dec		
	RD	45°	TD	RD	45°	TD
Yield strength (MPa)	166 ± 6.7	151 ± 1.7	150 ± 0.9	149 ± 0.1	138 ± 0.9	137 ± 0.3
Tensile strength (MPa)	258 ± 1.0	243 ± 8.0	224 ± 12.2	242 ± 1.6	235 ± 0.9	204 ± 23.8
Fracture strain (%)	25.0 ± 1.8	17.0 ± 5.2	8.9 ± 2.0	14.4 ± 0.6	12.2 ± 0.6	4.7 ± 2.5
Uniform elongation (%)	18.0 ± 1.1	15.0 ± 4.7	8.1 ± 1.6	13.9 ± 0.6	12.0 ± 0.5	4.6 ± 2.4

Figure 7 presents the Erichsen index (IE) of the as rolled and annealed sheet samples. The as-rolled sheets show very low formability, IE = 2.5 ~ 3, independently on the rolling scheme. After annealing for 600 s at 400 °C, improved formability, which is related to the reduction of lattice defects and the texture weakening, is achieved. A remarkable improvement of the formability is obtained in the 450-inc sheet after annealing. The average IE of 8.1 is almost a three-fold higher value in comparison to the commercial AZ31 sheet (IE = 2~3). Other sheets, rolled at 500 °C and 450-dec, have an IE of 4.0~4.5 after annealing. The excellent stretch formability of the 450-inc sheet after recrystallization annealing is attributed to the weak texture and the basal pole split into the RD and TD. As mentioned above, this texture type with a tetrarchy distribution of the basal poles is advantageous for the material flow in the sheet thickness direction during stretch forming due to the high activity of the basal <a> slip. Additionally, the fine grain structure with an average grain size of 6 μm and fine particles distributed homogeneously also enhance sheet formability. Considering that a large number of stringers of the secondary phase particles and Mg oxide aligning along the RD are found in all samples, we consider that the ductility in the TD and the stretch formability could be further improved. This requires the further optimization of the alloy composition as well as the improved casting technology. For instance, casting under vacuum or a controllable solidification rate ensures the reduced interaction between the melt and atmosphere.

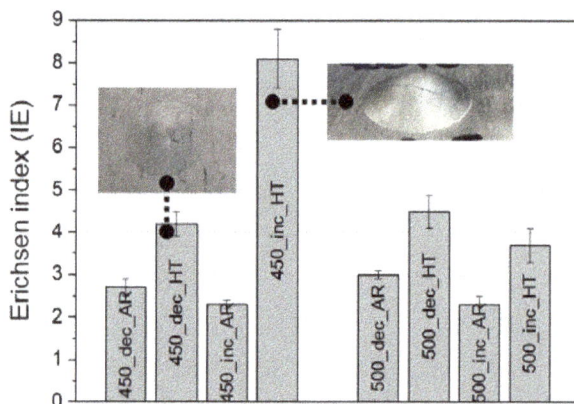

Figure 7. Erichsen index of the differently rolled and annealed sheets and appearances of the Erichsen samples of the 450-dec and 450-inc sheets (inset).

4. Summary

Non-flammable AZXW3100 alloy sheets were produced using various rolling conditions. Depending on the rolling schedule, distinct microstructure and texture evolution was observed during the warm rolling and subsequent recrystallization annealing.

The texture with tetrarchy characteristics, i.e., the basal pole split into TD and RD simultaneously, developed after rolling at 450 °C, with the deformation degree per pass increasing with the rolling step. The rolling at 450 °C with a decreasing deformation degree per pass resulted in a coarse grain structure with a relatively strong basal-type texture. After the sheet rolling at 500 °C, the texture with the basal pole split into the RD was developed, while the influence of the deformation degree per rolling step on the microstructure and texture became negligible.

The directional anisotropy of the mechanical properties is related to the distinct texture of each rolled sheet. It is obvious that the AZXW3100 sheet that has a weak texture with the basal pole split into the TD shows excellent sheet formability.

The present results clearly indicate that a highly ductile and formable sheet can be produced by microstructure and texture control, which in turn originate from optimized thermomechanical treatments. Further studies on deformation and recrystallization mechanisms in correlation with thermomechanical treatments will identify the fundamental mechanisms that produce such variety in microstructure and texture development.

Author Contributions: Experimental works were carried out by S.Y., J.V.-H. with continuous discussion with Y.M.K., B.S.Y. and D.L. All authors discussed the experimental results and the conclusions, and reviewed the present manuscript.

Acknowledgments: The authors appreciate A. Reichart and Y.K. Shin at HZG for technical support during the warm rolling and microstructure analysis. The financial support of the NST-National Research Council of Science & Technology (No. CRC-15-06-KIGAM) by the Korea government (MSIP) is gratefully acknowledged.

References

1. Bohlen, J.; Nürnberg, M.R.; Senn, J.W.; Letzig, D.; Agnew, S.R. The texture and anisotropy of magnesium-zinc-rare earth alloy sheets. *Acta Mater.* **2007**, *55*, 2101–2112. [CrossRef]
2. Stanford, N.; Barnett, M. Effect of composition on the texture and deformation behaviour of wrought Mg alloys. *Scr. Mater.* **2008**, *58*, 179–182. [CrossRef]
3. Chino, Y.; Kado, M.; Mabuchi, M. Enhancement of Tensile Ductility and Stretch Formability of Magnesium by Addition of 0.2 wt%(0.035 at%)Ce. *Mater. Sci. Eng.* **2008**, *A494*, 343–349. [CrossRef]

4. Bohlen, J.; Cano, G.; Drozdenko, D.; Dobron, P.; Kainer, K.U.; Gall, S.; Müller, S.; Letzig, D. Processing Effects on the Formability of Magnesium Alloy Sheets. *Metals* **2018**, *8*, 147. [CrossRef]

5. Tekumalla, S.; Seetharaman, S.; Almajid, A.; Gupta, M. Mechanical Properties of Magnesium-Rare Earth Alloy Systems: A Review. *Metals* **2015**, *5*, 1–39. [CrossRef]

6. Yi, S.; Bohlen, J.; Heinemann, F.; Letzig, D. Mechanical anisotropy and deep drawing behaviour of AZ31 and ZE10 magnesium alloy sheets. *Acta Mater.* **2010**, *58*, 592–605. [CrossRef]

7. Kim, Y.M.; Kim, H.S.; You, B.S.; Yim, C.D. Non-Flammable Magnesium Alloy with Excellent Mechanical Properties, and Preparation Method Thereof. U.S. Patent 2013/0183193 A1, 18 July 2013.

8. Gneiger, S.; Papenberg, N.; Frank, S.; Gradinger, R. *Investigations on Microstructure and Mechanical Properties of Non-Flammable Mg–Al–Zn–Ca–Y Alloys, Magnesium Technology 2018, TMS Annual Meeting & Exhibition, Phoenix, USA, March 11–15, 2018*; Orlov, D., Joshi, V., Solanki, K.N., Neelameggham, N.R., Eds.; Springer: Berlin/Heidelberg, Germany, 2018. [CrossRef]

9. Go, Y.; Jo, S.M.; Park, S.H.; Kim, H.S.; You, B.S.; Kim, Y.M. Microstructure and mechanical properties of non-flammable Mg-8Al-0.3Zn-0.1Mn-0.3Ca-0.2Y alloy subjected to low-temperature, low-speed extrusion. *J. Alloys Compd.* **2018**, *739*, 69–76. [CrossRef]

10. You, B.S.; Kim, Y.M.; Yim, C.D.; Kim, H.S. *Oxidation and Corrosion Behavior of Non-Flammable Magnesium Alloys Containing Ca and Y. Magnesium Technology 2014, TMS Annual Meeting & Exhibition, San Diego, USA, Feb. 16–20, 2014*; Alderman, M., Manuel, M.V., Hort, N., Neelameggham, N.R., Eds.; Springer: Berlin/Heidelberg, Germany, 2014. [CrossRef]

11. Stanford, N. The effect of calcium on the texture, microstructure and mechanical properties of extruded Mg–Mn–Ca alloys. *Mater. Sci. Eng.* **2010**, *A528*, 314–322. [CrossRef]

12. Kim, D.W.; Suh, B.C.; Shim, M.S.; Bae, J.H.; Kim, D.H.; Kim, N.J. Texture Evolution in Mg-Zn-Ca Alloy Sheets. *Metall. Mater. Trans.* **2013**, *A44*, 2950–2961. [CrossRef]

13. Chino, Y.; Huang, X.; Suzuki, K.; Mabuchi, M. Enhancement of Stretch Formability at Room Temperature by Addition of Ca in Mg-Zn Alloy. *Mater. Trans.* **2010**, *51*, 818–821. [CrossRef]

14. Bhattacharjee, T.; Suh, B.C.; Sasaki, T.T.; Ohkubo, T.; Kim Nack, J.; Hono, K. High strength and formable Mg–6.2Zn–0.5Zr–0.2Ca alloy sheet processed by twin roll casting. *Mater. Sci. Eng.* **2014**, *A609*, 154–160. [CrossRef]

15. Hofstetter, J.; Becker, M.; Martinelli, E.; Weinberg, A.M.; Mingler, B.; Kilian, H.; Pogatscher, S.; Uggowitzer, P.J.; Löffler, J.F. High-Strength Low-Alloy (HSLA) Mg–Zn–Ca Alloys with Excellent Biodegradation Performance. *JOM* **2014**, *66*, 566–572. [CrossRef]

16. Kree, V.; Bohlen, J.; Letzig, D.; Kainer, K.U. The metallographical examination of magnesium alloys. *Pract. Metallogr.* **2004**, *41*, 233–246.

17. MTEX Toolbox. Available online: http://mtex-toolbox.github.io/ (accessed on 28 December 2018).

18. Agnew, S.R.; Yoo, M.H.; Tomé, C.N. Application of texture simulation to understanding mechanical behavior of Mg and solid solution alloys containing Li or Y. *Acta Mater.* **2001**, *49*, 4277–4289. [CrossRef]

19. Al-Samman, T.; Gottstein, G. Room temperature formability of a magnesium AZ31 alloy: Examining the role of texture on the deformation mechanisms. *Mater. Sci. Eng.* **2008**, *A488*, 406–414. [CrossRef]

20. Hantzsche, K.; Bohlen, J.; Wendt, J.; Kainer, K.U.; Yi, S.B.; Letzig, D. Effect of rare earth additions on microstructure and texture development of magnesium alloy sheets. *Scr. Mater.* **2010**, *63*, 725–730. [CrossRef]

21. Stutz, L.; Bohlen, J.; Kurz, G.; Letzig, D.; Kainer, K.U. Influence of the processing of magnesium alloys AZ31 and ZE10 on the sheet formability at elevated temperature. *Key Eng. Mater.* **2011**, *473*, 335–342. [CrossRef]

22. Suh, J.; Victoria-Hernández, J.; Letzig, D.; Golle, R. Effect of processing route on texture and cold formability of AZ31 alloy sheets processed by ECAP. *Mater. Sci. Eng.* **2016**, *A669*, 159–170. [CrossRef]

metals

MDPI

Article

Effect of Surface Roughness on the Bonding Strength and Spring-Back of a CFRP/CR980 Hybrid Composite

Ji Hoon Hwang [1], Chul Kyu Jin [2], Min Sik Lee [1], Su Won Choi [1] and Chung Gil Kang [3,*]

[1] Precision Manufacturing System Division, Graduate School, Pusan National University, San 30 Chang Jun-dong, Geum Jung-Gu, Busan 46241, Korea; hoonida_731@naver.com (J.H.H.); minsik2@pusan.ac.kr (M.S.L.); suwon9180@naver.com (S.W.C.)
[2] School of Mechanical Engineering, Kyungnam University, 7 Kyungnamdaehak-ro, Masanhappo-gu, Changwon-si 51767, Korea; cool3243@kyungnam.ac.kr
[3] School of Mechanical Engineering, Pusan National University, San 30 Chang Jun-dong, Geum Jung-Gu, Busan 46241, Korea
* Correspondence: cgkang@pusan.ac.kr; Tel.: +82-51-510-1455

Received: 12 July 2018; Accepted: 10 September 2018; Published: 12 September 2018

Abstract: Carbon fiber-reinforced plastic (CFRP), which is a light and composite material, has a higher specific strength and stiffness than metal materials. However, owing to its low elongation, it is vulnerable to local impacts such as collision. Therefore, hybrid composite materials that can overcome the disadvantages of homogeneous materials by bonding CFRP and metal materials are increasingly popular. In this study, a physical surface treatment sandblast process was applied on a high tensile steel plate (CR980) manufactured by cold rolling to form another surface condition, and the bonding strength with CFRP was measured. In addition, spring-back due to the manufacturing process of the CFRP and CR980 hybrid composite material bonded with different surface roughness was observed. The bonding strength and the spring-back angle of the CFRP/CR980 hybrid composite material tended to increase with the increase in the surface roughness.

Keywords: hybrid composite material; V-bending test; spring-back; surface roughness; shear lap test

1. Introduction

In the case of eco-friendly automobiles, additional components like batteries and motors are added for the development of new power sources, which has lead inevitably to an increase in the vehicle weight. On the basis of the European 2020 target, the use of non-ferrous metals and synthetic resins in eco-friendly vehicles will each increase by more than 10% [1].

Recently, the demand for the comfort and safety of automobiles, such as body reinforcements, airbags, automobile electric motors, and accumulators, has increased, and the weight of the vehicle is showing an increasing trend. Therefore, the problem of automobile weight should be addressed [2].

In general, the performance improvement of automobile by a 10% reduction in the weight of the automobile body contributes a 6–8% improvement in fuel efficiency, 8% improvement in acceleration and braking performance, 6% improvement in steering performance, 4.5% reduction in CO, 4.5% reduction in HC, and 8.8% reduction in NO_x.

Recently, aluminum alloys, magnesium alloys, titanium alloys, advanced high strength steel (AHSS), and composite materials are increasingly being used in the automobile industry. Many studies have investigated the use of carbon fiber-reinforced plastic (CFRP), which is a composite material, owing to its higher specific strength and stiffness than steel material [3–5]. In addition, studies have been carried out on the tensile failure mechanism of CFRP prepreg (pre-impregnated composite fibers) [6] and comparison of the tensile strengths of the different epoxy composite materials of CFRP [7]. Prepreg is a material in which matrix epoxy resin is impregnated in the carbon fiber, which is an intermediate material of CFRP as well as a reinforcement material. CFRP is used in a variety of fields, such as aircraft, leisure goods, and automobiles, owing to its lightweight properties. However, despite its high tensile strength, CFRP is vulnerable to local impacts like collisions, owing to its low elongation (approximately 2%). In recent years, there has been increased research interest on hybrid composites that overcome the limitations of materials, by bonding CFRP and metal materials to address the problems of such a heterogeneous material. Studies on interfacial bonding between CFRP and steel [8] and studies on epoxy flow in deep drawing processes in CR340 and CFRP composites have been carried out [9,10].

As early as 1879, Thomas Edison discovered that one could bake cotton and bamboo bioderived materials at high temperatures, resulting in carbonization into a carbon fiber filament. This was used in the first lightbulb of an incandescent nature that is powered by electricity [11]. As such, materials from nature should be considered when understanding the spring-back response, due to the viscoelastic nature of the cellulose polymer embedded around a sheath of protective lignin and bonded hemicellulose for stress transfer [12,13]. The significant response in wood-based composites is a classic example of differential swell in different components of natural materials that results in delamination [14]. As such, it is anticipated that carbon fibers may be subject to spring-back, due to the specific elasticity of a polymer originally designed by nature.

In the case of CFRP and metal hybrid composites, studies on spring-back are lacking. Therefore, research on spring-back after forming CFRP and metal hybrid composite material is necessary. In this research, changes in spring-back according to changes in the production process was investigated through a primary experiment, such as V-being before manufacturing the parts with complicated shapes (like automobile parts) considering the lamination sequence of the CFRP, the number of lamination prepreg, and lamination direction.

CR980 material utilized as automobile material as used in this study is excellent in terms of not only its economic feasibility and light weight, but also its high strength compared with alternative materials, such as aluminum. In particular, CR980 has an excellent process capability, such that products having very complicated shapes can be produced compared with using aluminum. With the application of the hot press forming process followed by the expansion of high-strength steel, the need for new process technology to solve process factors with a low feasibility is increasing due to an increase in the mold processing cost, installation cost of development equipment, and cycle time. The steel used in this study was a high tensile steel plate (CR980) manufactured by cold rolling. CR980 material is a recently developed advanced high strength steel, which is widely used as a pillar part and the underframe of a car body, which plays very important roles by absorbing impacts during collision of cars. A lap shear test was carried out using a hybrid material mechanically bonded according to variations in the degree of surface treatment during bonding of the two materials, CFRP and CR980, with the process parameters of surface roughness, compressive force, and compression direction. Spring-back measurements for the specimen having V-bending according to the radius of the punch and die was also carried out.

In the case of a hybrid composite material, the reinforcement material and matrix may not be well bonded. In this case, delamination occurs, which impairs the absorption capability. To improve the bonding strength of these heterogeneous materials, physical surface treatment was carried out using the sandblast method, which is a physical surface treatment, and the degree of surface roughness was measured by the Rz value of a 10-point average roughness. Experiments were conducted to investigate the change in the bond strength according to the degree of surface roughness and variation in the bonding pressure.

2. Experiments

2.1. Shear Lap Adhesion Test of the CFRP/CR980 Hybrid Composite

2.1.1. Preparation of Specimen

Figure 1 shows a schematic diagram of the equipment for fabricating CFRP using 10 prepregs.
An optimum condition to fabricate a CFRP involved compressing ten sheets of prepreg at 140 °C
at a pressure of 0.5 MPa. The curing of the epoxy was performed for 30 min [15]. The CFRP used
in the experiment was a plain carbon fiber from TORAY. The thickness of the prepreg was 0.27 mm,
and the thermosetting prepreg had an initial epoxy weight percentage of 42 wt %. CR980 steel is a
material that is applied to automobile pillars and underframes, which generally protects the driver
from external impact. The thickness of the CR980 material used was 1.2 mm. For 0 degrees of rolling
direction, the tensile strength and elongation were 900 MPa and 18.8%, respectively. For 90 degrees of
rolling direction, the tensile strength and elongation were 1053 MPa and 18.0%, respectively. Values of
normal anisotropic (R) for 0 and 90 degrees are 0.57 and 0.65, respectively. Details of the material are
presented in Tables 1 and 2.

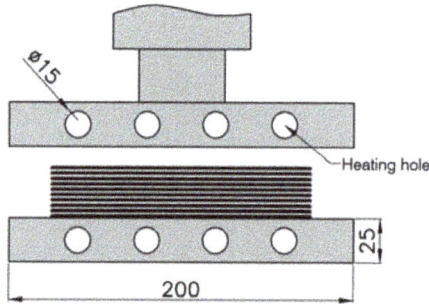

Figure 1. Schematic diagram of the equipment for fabricating carbon fiber-reinforced plastic (CFRP)
using 10 pre-impregnated composite fibers (prepregs) (unit: mm).

Table 1. Mechanical properties of CFRP prepreg (CF3327EPC, HANKUK CARBON) [16].

Construction	Weight of Carbon Fiber	Weight of Resin	Resin Content	Total Weight	Fabric Thickness
Plain	205 g/m^2	150 gr/m^2	42 ± 2%	352 gr/m^2	0.27 ± 0.05 mm

Table 2. Mechanical Properties of carbon fiber (T300-3K-50, TORAY INDUSTRIES Inc., Tokyo, Japan)
and CR980 Steel (SPFC980Y, HYUNDAI STEEL, Seoul, Korea) [17,18].

Construction		Yield Strength	Tensile Strength	Elongation	Elastic Modulus
3K Carbon fiber			3530 MPa	1.5%	E1 = 135 GPa, E2 = 10 GPa
SPFC980Y	(RD0°)	568 MPa	900 MPa	18.8%	200 GPa
	(RD90°)	619 MPa	1043 MPa	18.0%	

2.1.2. Production of Specimen

A lap shear adhesion test was conducted to measure the bonding strength. The test specimen
production method is shown in Figure 2. Hot compression molding process was used to produce test
specimens for bonding strength measurement. Table 3 lists the parameters of the hot compression
molding process.

Table 3. Parameters of the hot compression molding process.

Surface Roughness (Rz) between CFRP and CR980	Pressure	Temperature of Mold	Curring Time
20, 35, 45, 60 μm	1, 2 MPa	160 °C	30 min

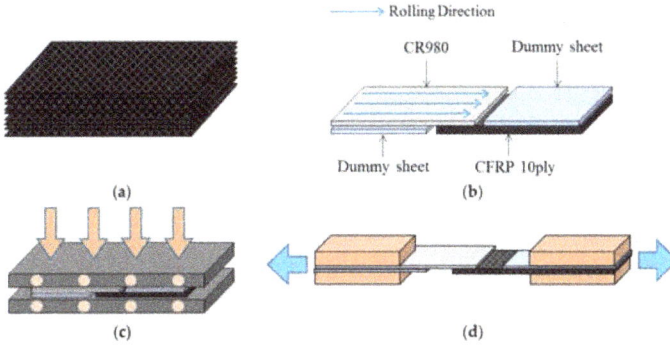

Figure 2. A method of making test specimens for bonding strength measurements, and a schematic diagram of shear testing: (**a**) CFRP 10-ply; (**b**) CFRP/CR980 and dummy sheet laminating; (**c**) hot compression molding; (**d**) shear lap adhesion test.

Sandblast surface treatment was carried out on the surface of one side. Sandblasting is a process to give the surface of CR980 a physically rough surface by spraying sand with compressed air. The degree of roughness is categorized according to the particle size of the sand. Unnecessary deposits are removed using sandblast, and the surface of the CR980 becomes a fine uneven surface, making the epoxy more permeable, in order to measure the difference in the bonding strength. Ten CFRP specimens were produced by laminating CR980 materials, each having a surface roughness of Rz = 20 μm, 35 μm, 45 μm, and 60 μm, respectively. The sizes of the specimens were 100 × 25 mm. The bonding area of the two materials was set to 20 × 25 mm at the center, and pressed and heated at 160 °C at 1 MPa and 2 MPa of pressure, respectively, followed by curing for 30 min. For precise temperature control, the room temperature was kept constant and the temperature of each mold was measured and controlled using a heat controller and heat cartridge. The equipment used in the experiment was a 25-ton Material Testing System (MTS). For this bonding strength measurement experiment, an appropriate manufacturing method was used by referring to the KS M 3713: 2012 standard [19]. The overall test method of the CFRP is also presented in ASTM D4762-16 [20].

Figure 3 show photographs of the preparation of the test specimen using the above method. Figure 4 show photographs of CR980 material that has been surface treated with sandblast, whose surface roughness was measured using an optical roughness tester. Figure 4a shows the specimen of surface treatment condition of Rz = 20 μm, Figure 4b shows the surface treatment condition of Rz = 35 μm, Figure 4c shows a photograph of the specimen with the surface treatment condition Rz = 45 μm, and Figure 4d shows photographs of surface roughness with the treatment condition Rz = 60 μm.

The shear lap adhesion test for the bonding strength was carried out by setting the shear rate to 2 mm/min. The steel tab was attached to the grip to prevent the grip of the specimen from being damaged by the bite of the machine.

Figure 3. Fabricated specimen for shear lap adhesion test.

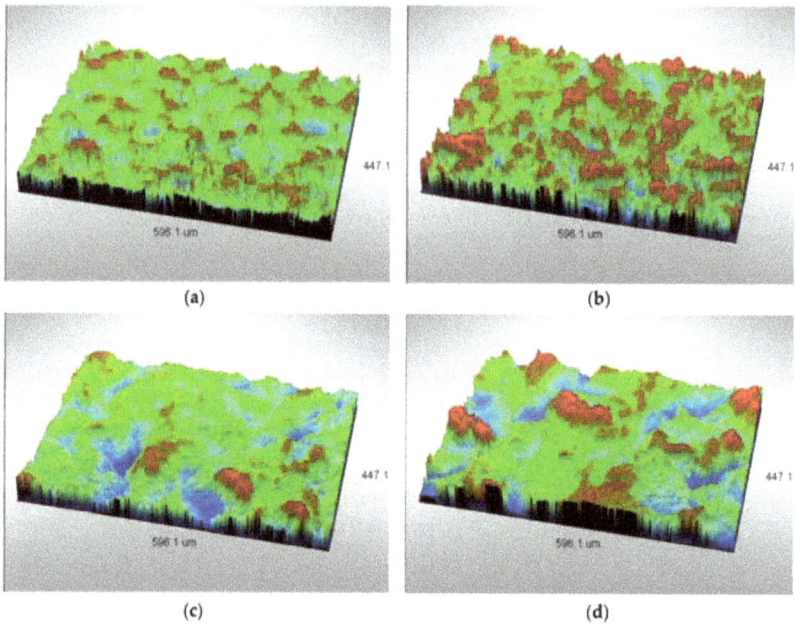

Figure 4. Surface roughness images of CR980 conducted by sandblast: (**a**) Rz = 20 μm; (**b**) Rz = 35 μm; (**c**) Rz = 45 μm; (**d**) Rz = 60 μm.

2.2. V-Bending Test of CFRP/CR980 Hybrid Composite

In order to measure spring-back of the CFRP/CR980 hybrid composite, V-bending experiments were conducted. Figure 5 shows a detailed view of the shape of the mold used in the V-bending test of the CFRP/CR980 hybrid composite. The mold was divided into an upper punch part and a lower die part. To control the temperature of the punch part and the die part, a hole was made for the heat cartridge to be inserted. The edge part of the die and punch have radius (R) of 5 mm. Table 4 lists the parameters of the V-bending test of the CFRP/CR980 hybrid composite. Experiments were conducted under 1 MPa and 2 MPa, respectively. The die temperature of 160 °C was maintained. After the V-bending specimen has been pressed and cured using constant pressure for 30 min, the V-bending angle was measured using a digital protractor from BLUEBIRD.

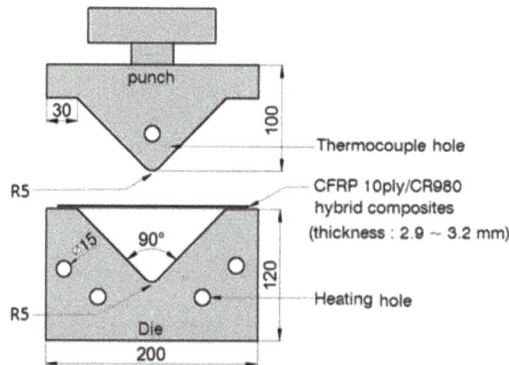

Figure 5. Drawing of mold for V-bending test of CFRP/CR980 hybrid composites (unit: mm).

Table 4. Parameters of the V-bending test of the CFRP/CR980 hybrid composite.

Pressure	Temperature of Mold	Curring Time	Direction of CR980 Sheet
1, 2 MPa	160 °C	30 min	RD0°, RD90°

In case of the CR980 material produced by cold rolling, there is a difference in the mechanical property value and the elastic recovery amount according to the rolling direction. When V-bending was carried out in the direction parallel to the rolling direction, it was indicated as RD0°, while when V-bending was carried out in the vertical direction, it was indicated as RD90°. The concept of the rolling direction and V-bending is shown in Figure 6. The size of the specimen used in the V-bending experiment was set at 180 × 130 mm. The equipment used in the experiment was a 25-ton MTS.

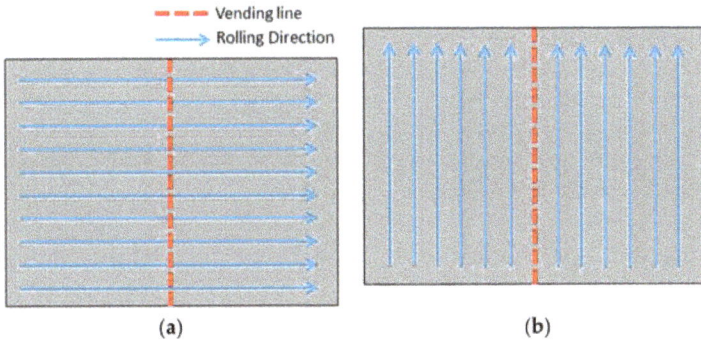

Figure 6. Rolling direction of CR980 for V-bending test: (**a**) 90° (RD90°); (**b**) 0° (RD0°).

Figure 7 shows a diagram of a spring-go phenomenon occurring in a direction opposite to that of a spring-back phenomenon, in accordance with the elastic recovery amount after the bending process of the plate material. In general, during V-bending by physical pressure, compressive stress occurs in the plate material at the punch portion of the mold, while tensile stress occurs at the die portion. As the punch ascends, the punch portion where the compressive stress occurs is restored to its original shape; conversely, the spring-back phenomenon occurs in the die portion, where the tensile stress occurs due to contraction.

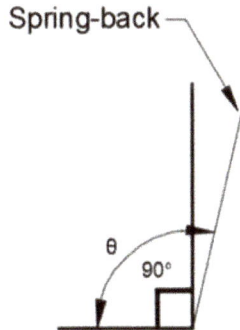

Figure 7. A conceptual view of the spring-back.

3. Results

3.1. CFRP/CR980 Hybrid Composite by Shear Lap Adhesion Test

3.1.1. Bonding Strength

Figure 8 shows the bonding strength of CFRP/CR980 hybrid composites with different surface roughness conducted by the hot compression molding process under 1 MPa and 2 MPa of pressure. When the pressure was 1 MPa, the average bonding strength of the control was measured as 3.7 MPa. The bonding strengths were measured as 10.2 MPa, 15.2 MPa, 17.1 MPa, and 10.2 MPa when the surface roughness values were 20 μm, 35 μm, 45 μm, and 60 μm, respectively. When the pressing condition was 2 MPa, the average bonding strength of the untreated specimens was measured as 4.1 MPa, while when the surface roughness values were 20 μm, 35 μm, 45 μm, and 60 μm, and the bonding strengths were measured as 7.5 MPa, 9.2 MPa, 10.5 MPa, and 9.2 MPa, respectively.

As the surface roughness of the CR980 increased, it was found that the bonding strength of the CFRP/CR980 hybrid composite also increased. However, the bond strength decreased at the roughest surface of 60 μm. The bonding strength under 2 MPa of the pressure was lower than that under 1 MPa of the pressure. A possible reason for the decrease of the bonding strength could be the loss of epoxy, which is the matrix of the CFRP prepreg, and the largest factor in the bonding characteristics with increasing pressure applied to the specimens.

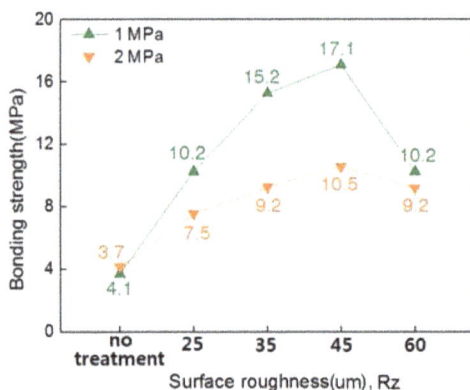

Figure 8. Bonding strength of CFRP/CR980 hybrid composites with different surface roughness conducted by hot compressing under different pressures.

Figure 9 shows the specimens after Shear Lap adhesion test. Figure 9a is the specimen without sandblast treatment conducted by 1 MPa of the pressure and Figure 9b is the specimen with surface roughness (Rz = 45 μm) conducted by 1 MPa of the pressure. Fracturing from the shear lap adhesion test occurred at about 12 mm of the bonding area. In the case of a specimen with surface roughness, there was some epoxy and CFRP on the surface of CR980. On the other hand, the CR980 specimen without sandblast treatment had no epoxy and CFRP on the surface.

(a)   (b)

Figure 9. Specimens after the shear lap adhesion test (bonding pressure: 1.0 MPa): (a) without sandblast treatment; (b) with surface roughness (Rz = 45 μm).

3.1.2. Microstructures

As can be seen from the bonding strength measurement experiment, the bonding strength of the CFRP/CR980 hybrid composite increases as the surface roughness of the CR980 increases. However, at Rz = 60 μm, which is the roughest surface, the bonding strength decreased under the condition of pressure forces of 1 MPa and 2 MPa.

To clarify the reason for the decrease in the bonding strength, photographs of the cross-section of the specimen with Rz = 45 μm, which is the surface roughness of CR980 with the highest bond strength, and the specimen with a surface roughness of 60 μm, whose bond strength decreased, were observed.

Figure 10 show photographs of the microstructure of the cross-section of the CFRP/CR980 hybrid composites. Figure 10a show the specimen with a surface roughness of 45 μm, while Figure 10b show the specimens with a surface roughness of 60 μm. In the specimen with the highest bonding strength with surface roughness Rz = 45 μm, dark gray epoxy was evenly distributed at the interface of the bonding of CFRP and CR980, indicating that the epoxy effect the increase in bonding strength. The specimen with Rz = 60 μm shows the roughest surface treatment, which means that the surface roughness increased compared with the specimen with Rz = 45 μm. As the surface roughness increases, epoxy, which affects the bonding strength, cannot be uniformly impregnated, and pores are generated. It can be observed that the bonding strength of the specimen having the roughest surface (Rz = 60 μm) decreased.

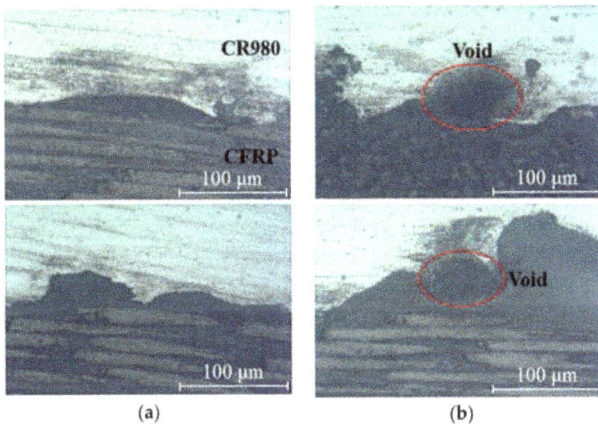

Figure 10. Microstructures (section view) CFRP/CR980 hybrid composites conducted by shear lap adhesion test: (**a**) surface roughness Rz = 45 μm; (**b**) surface roughness Rz = 60 μm.

3.2. CFRP/CR980 Hybrid Composite by V-Bending Test

3.2.1. Thickness

Figure 11 shows the thickness of the CFRP/CR980 hybrid composite conducted by V-bending at different positions. When the number of laminations of the CFRP prepreg was 10, the thickness was in the range of 2.85–3.11 mm. In general, a thickness of 2–3 mm is suitable for use in automotive pillar and subframe components. To meet these requirements, the use of CR980 1.2 mm material and 10 sheets of CFRP prepreg to reduce the overall thickness of the latter half part from 2 mm to approximately 3 mm, is proposed as the most suitable layer for the part. In addition, the thickness of the CFRP/CR980 hybrid composite tends to be thicker at the side of the specimen than at the center of the specimen, which could be due to the concentration of pressure at the center than at the side of the punch and die.

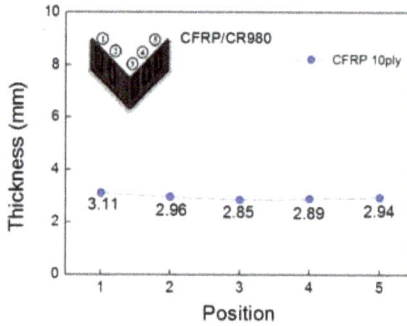

Figure 11. The thickness of the CFRP/CR980 conducted by V-bending test at different positions.

3.2.2. Spring-Back Angle

The punch with a V-shape loading, puts a concentrated load on CFRP/CR980 hybrid composite and does get dispersed evenly before causing any damage. Figure 12 shows the spring-back angle of the CFRP/CR980 hybrid composite with different surface roughness conducted by a V-bending test under 1 MPa of the pressure. As the surface roughness values increased to 20 μm, 35 μm, 45 μm, and 60 μm, when the number of lamination of CFRP was 10, the rolling direction was 0° (RD0°) and the pressure force was 1 MPa, the spring-back angles were 97.8°, 98.0°, 99.6°, 99.9°, and 98.9°, respectively. As the surface roughness increased, the spring-back angle also increased. Similarly, when the rolling direction was 90° (RD90°), the spring-back angles were 98.0°, 99.7°, 100°, 100.3°, and 99.9°, respectively. The spring-back angle was larger in the rolling direction of 90° (RD90°) than 0° (RD0°).

Figure 13 shows the spring-back angle of the CFRP/CR980 hybrid composite with different surface roughness conducted by a V-bending test under 2 MPa of the pressure. When the number of laminations of the CFRP was 10, the rolling direction was 0° (RD0°) and pressure force was 2 MPa; then, as the surface treatment condition increased without treatment to Rz = 20 μm, 35 μm, 45 μm, and 60 μm, the spring-back angle increased to 97.6°, 98.0°, 99.6°, 100.0°, and 100.0°, respectively. Overall, as the surface roughness increased, the spring-back also increased. Similarly, when the rolling direction was 90° (RD90°), the spring-back also increased. The spring-back angle was larger when the rolling direction was 90° (RD90°) than when it was 0° (RD0°). Choi et al. present that the spring-back angles were 93.9° (RD0°) and 93.7° (RD90°) at the V-bending test of CFRP 10ply/CR340 without treatment under 0.5 MPa, 140 °C, and 30 min of curing [21]. The spring-back angles of CFRP/CR980 were slightly higher than CFRP/CR340, though the process parameters were a little different.

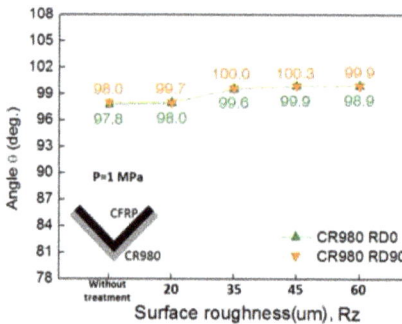

Figure 12. Angle of spring-back of CFRP/CR980 hybrid composites with different levels of surface roughness, conducted by a V-bending test under 1 MPa of the pressure.

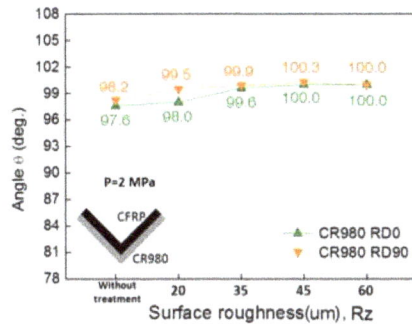

Figure 13. Angle of spring-back of CFRP/CR980 hybrid composites with different levels of surface roughness, conducted by a V-bending test under 2 MPa of the pressure.

3.2.3. Microstructures

Figure 14 shows microstructures at different positions of the CFRP/CR980 hybrid composite, conducted by a V-bending test under 1 MPa of the pressure. Position ① shows the nearest area of curvature of the punch and die part. As a result, it can be observed at position ① that delamination occurred at the bonding interface of the CFRP and CR980 without treatment. In case of the conditions of Rz = 20 μm, 35 μm, 45 μm, and 60 μm, no delamination was observed. However, under the Rz = 60 μm condition, it can be considered that pores were formed because epoxy, which is a matrix of the CFRP prepreg, could not be uniformly impregnated on the surface. Position ② shows the middle area of the CFRP/CR980 hybrid composite. Delamination at the interface of the CFRP and CR980 did not occur at position ②. Position ③ shows the edge area of the CFRP/CR980 hybrid composite. No delamination phenomenon was observed, despite changes in the roughness of the CR980 and without treatment conditions. However, the condition of Rz = 60 μm shows that epoxy was not uniformly impregnated on the surface of the CR980, and pores were observed. The interface bond strength between the matrix and CFRP prepreg had a significant effect on the surface roughness.

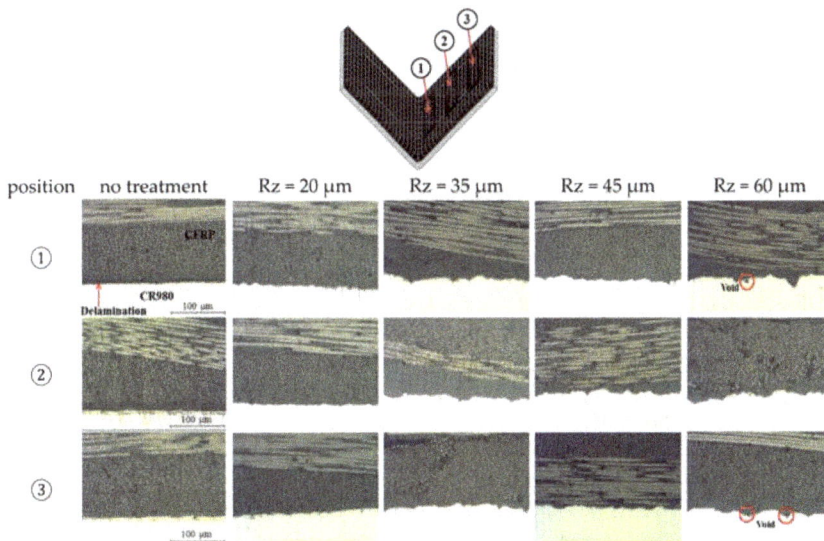

Figure 14. Microstructures (section view) at different positions of CFRP/CR980 hybrid composites conducted by a V-bending test.

4. Conclusions

In this study, the bonding strength and the spring-back of the CFRP/CR980 hybrid composite was investigated:

(1) The bonding strength between the surface-treated CR980 and CFRP generally increased depending on the degree of surface treatment. In the condition of Rz = 60 μm or more, the bonding strength was reduced.

(2) The bonding strength of specimens pressurized at 2 MPa was lower than that at 1 MPa, when pressure was applied during the production of shear test specimen to measure the bond strength.

(3) The spring-back angle tended to increase with an increase in the surface roughness. There is a little bit of difference in spring-back angles between 1 MPa and 2 MPa of the pressure.

(4) Delamination occurred at the bonding interface near the punch contact area of the CFRP/CR980 hybrid composite without treatment under 1 MPa of the pressure.

Author Contributions: J.H.H. and C.K.J. designed the experiment tools and performed the experiment. C.K.J., M.S.L., and S.W.C. analysed the experimental results, whereas Chung Gil Kang maintained and examined them. All authors have contributed to the discussions as well as revisions.

Acknowledgments: This work was supported by the National Research Foundation of Korea (NRF) grant funded by the Korean government (MSIP) through the GCRC-SOP (No. 2011-0030013). This work was also supported by the National Research Foundation of Korea (NRF) grant funded by the Korea government (MSIT) (No. 2017R1A2B4007884). This work was also supported by the National Research Foundation of Korea Grant funded by the Korean Government (NRF-2017R1C1B5017242).

Conflicts of Interest: The authors declare no conflict of interest.

References

1. Lightweighting-Wikipedia. Available online: https://en.wikipedia.org/wiki/Lightweighting (accessed on 1 September 2018).
2. Lightweight-Materials-Cars-and-Trucks/Department of Energy. Available online: https://www.energy.gov/eere/vehicles/lightweight-materials-cars-and-trucks (accessed on 1 September 2018).
3. Al-Zubaidy, H.; Zhao, X.L.; Al-Mihaidi, R. Mechanical Behaviour of Normal Modulus Carbon Fibre Reinforced Polymer (CFRP) and Epoxy under Impact Tensile Loads. *Procedia Eng.* **2011**, *10*, 2453–2458. [CrossRef]
4. Van-Paepegem, W.; De-Geyter, K.; Vanhooymissen, P.; Degrieck, J. Effect of friction on the hysteresis loops from three-point bending fatigue tests of fibre-reinforced composites. *Compos. Struct.* **2006**, *72*, 212–217. [CrossRef]
5. Kleiner, M.; Geiger, M.; Klaus, A. Manufacturing of Lightweight Components by Metal Forming. *CIRP Ann.-Manuf. Technol.* **2003**, *52*, 521–542. [CrossRef]
6. Fuwa, M.; Bunsell, A.R.; Harris, B. Tensile failure mechanisms in carbon fibre reinforced plastics. *J. Mater. Sci.* **1975**, *10*, 2062–2070. [CrossRef]
7. Paiva, J.M.F.; Mayer, S.; Rezende, M.C. Comparison of Tensile Strength of Different Carbon Fabric Reinforced Epoxy Composites. *Mater. Res.* **2006**, *9*, 83–89. [CrossRef]
8. Yu, T.; Fernando, D.; Teng, J.G.; Zhao, X.L. Experimental study on CFRP-to-steel bonded interfaces. *Compos. Part B Eng.* **2012**, *43*, 2279–2289. [CrossRef]
9. Lee, M.S.; Kim, S.J.; Lim, O.D.; Kang, C.G. Effect of process parameters on epoxy flow behavior and formability with CR340/CFRP composites by different laminating in deep drawing process. *Procedia Eng.* **2014**, *81*, 1627–1632. [CrossRef]
10. Lee, M.S.; Kim, S.J.; Kim, H.H.; Lim, O.D.; Kang, C.G. Effects of process parameters on epoxy flow behavior and formability in deep drawing process with CR340/carbon fiber–reinforced plastic composites. *Proc. Inst. Mech. Eng. B J. Eng. Manuf.* **2014**, *229*, 86–99. [CrossRef]
11. High Performance Carbon Fibers—National Historic Chemical Landmark. Available online: https://www.acs.org/content/acs/en/education/whatischemistry/landmarks/carbonfibers.html (accessed on 1 September 2018).

12. Fan, L.-T.; Gharpuray, M.M.; Lee, Y.H. Nature of Cellulosic Material. *Cellul. Hydrolys.* **1987**, *3*, 5–20.

13. Via, B.K.; So, C.L.; Shupe, T.F.; Groom, L.H.; Wikaira, J. Mechanical response of longleaf pine to variation in microfibril angle, chemistry associated wavelengths, density, and radial position. *Compos. Part A Appl. Sci. Manuf.* **2009**, *40*, 60–66. [CrossRef]

14. Wei, P.; Rao, X.; Yang, J.; Guo, Y.; Chen, H.; Zhang, Y.; Wang, Z. Hot pressing of wood-based composites: A review. *For. Prod. J.* **2016**, *66*, 419–427. [CrossRef]

15. Lee, M.S.; Kim, S.J.; Lim, O.D.; Kang, C.G. A study on mechanical properties of Al5052/CFRP/Al5052 composite through three-point bending tests and shear lap tests according to surface roughness. *J. Compos. Mater.* **2016**, *10*, 1–11. [CrossRef]

16. Korea Carbon. Available online: https://www.hcarbon.com/product/overview.asp (accessed on 1 September 2018).

17. Toray. Available online: https://www.toray.com/products/prod_001.html (accessed on 1 September 2018).

18. Hyundai Steel. Available online: https://www.hyundai-steel.com/kr/products-technology/products/hotrolledsteel.hds (accessed on 1 September 2018).

19. Korean Agency for Technology and Standard. *Methods of Making Samples of Carbon Fiber Reinforced Plastics*; KS M 3713; KATS: Seoul, Korea, 2012.

20. *Standard Guide for Testing Polymer Matrix Composite Materials*; ASTM D4762-16; ASTM International: West Conshohocken, PA, USA, 2011.

21. Choi, S.W.; Lee, M.S.; Kang, C.G. Effect of process parameters and laminating methods on spring-back in V-bending of CFRP/CR340 hybrid composites. *Int. J. Precis. Eng. Manuf.* **2016**, *17*, 395–400. [CrossRef]

metals

MDPI

Article

Connected Process Design for Hot Working of a Creep-Resistant Mg–4Al–2Ba–2Ca Alloy (ABaX422)

Kamineni Pitcheswara Rao [1,*], Dharmendra Chalasani [1,†], Kalidass Suresh [1,‡], Yellapregada Venkata Rama Krishna Prasad [2], Hajo Dieringa [3,*] and Norbert Hort [3]

1   Department of Mechanical and Biomedical Engineering, City University of Hong Kong, Tat Chee Avenue, Kowloon, Hong Kong 999077, China; dharmendra.chalasani@unb.ca (D.C.); ksureshphy@buc.edu.in (K.S.)
2   Independent Researcher (formerly with City University of Hong Kong), No. 2/B, Vinayaka Nagar, Hebbal, Bengaluru 560024, India; prasad_yvrk@hotmail.com
3   Magnesium Innovation Centre, Helmholtz Zentrum Geesthacht, Max-Planck-Strasse 1, 21502 Geesthacht, Germany; norbert.hort@hzg.de
*   Correspondence: mekprao@cityu.edu.hk (K.P.R.); hajo.dieringa@hzg.de (H.D.); Tel.: +852-3442-8409 (K.P.R.); Fax: +852-3442-0172 (K.P.R.)
†   Current address: Department of Mechanical Engineering, University of New Brunswick, Fredericton E3B5A1, NB, Canada.
‡   Current address: Department of Physics, Bharathiar University, Coimbatore 641046, India.

Received: 12 May 2018; Accepted: 13 June 2018; Published: 18 June 2018

Abstract: With a view to design connected processing steps for the manufacturing of components, the hot working behavior of the ABaX422 alloy has been characterized for the as-cast and extruded conditions. In the as-cast condition, the alloy has a limited workability, due to the presence of a large volume of intermetallic phases at the grain boundaries, and is not suitable to process at high speeds. A connected processing step has been designed on the basis of the results of the processing map for the as-cast alloy, and this step involves the extrusion of the cast billet to obtain a 12 mm diameter rod product at a billet temperature of 390 °C and at a ram speed of 1 mm s^{-1}. The microstructure of the extruded rod has a finer grain size, with redistributed fine particles of the intermetallic phases. The processing map of the extruded rod exhibited two new domains, and the one in the temperature range 360–420 °C and strain rate range 0.2–10 s^{-1} is useful for manufacturing at high speeds, while the lower temperature develops a finer grain size in the product to improve the room temperature strength and ductility. The area of the flow instability is also reduced by the extrusion step, widening the workability window.

Keywords: Mg-Al-Ba-Ca alloy; microstructure; strength; hot working; kinetic analysis; processing map

1. Introduction

High temperature creep resistance is an important requirement for Mg alloys for automobile and aerospace components [1]. From this viewpoint, commercial Mg alloys like AE42 and MRI230D have been developed [2,3], in which creep strength is improved by the addition of rare-earth and alkaline-earth elements like Ca and Sr, respectively. In Mg–Al and Mg–Al–Zn alloys, the room temperature and elevated temperature strength are improved by the addition of Ca and Sr [4–6] through microstructural refinement and texture strengthening [7,8]. However, the quest for finding better creep-resistant Mg alloys continues, and a new alloy class is based on an Mg–Al–Ba–Ca (ABaX) system [9–11], where the strength is increased with increasingly higher alloying content. Starting from ABaX421, the percentages of Al, Ba, and Ca have been successively increased to develop ABaX422, ABaX633, and ABaX844 alloys [12–16]. The strength of these alloys is not only better than the conventional heat-resistant commercial Mg alloys, but has also increased with increasing alloying content, in the order given above. These alloys have all been studied in the as-cast condition, in which the microstructure has a large volume fraction of coarse $Mg_{21}Al_3Ba_2$ and $(Mg,Al)_2Ca$ intermetallic phases at the grain boundaries, which are stable up to the melting point of the alloy [10]. The higher concentration alloys exhibit low ductility and workability, due to the coarseness of the cast structure and intense chemical segregation, and generally require slow speeds for hot working [12–16], making the manufacturing of hot worked components unviable. However, the constitutive response that decides the hot workability is sensitive to the prior processing history [17], among other variables, and therefore attempts may be made to change the response of the material by carefully designing connected processing steps. To achieve this goal, the technique of processing maps [17–19] is a powerful tool, since it is capable of predicting the constitutive response of the material without the trial and error procedure. The aim of this investigation is to design connected processing steps for manufacturing products like forgings, extrusions, and rolling stock of the ABaX422 alloy, starting from cast billet. In this process design, the optimum parameters will be selected on the basis of the predictions of the processing map with regard to the microstructural mechanisms that dissipate power during hot working.

Processing maps are developed based on the principles of the dynamic materials model, details of which are described in earlier studies [17–19]. Briefly, the model considers the workpiece undergoing hot deformation as a non-linear dissipater of power, which occurs in the form of two complementary parts: mainly in the form of deformation heat (G content), and as dissipative microstructural mechanisms (J co-content). The factor that partitions the power is the strain rate sensitivity (m) of the flow stress [17–19]:

$$\frac{dJ}{dG} = \frac{\dot{\varepsilon}\,d\bar{\sigma}}{\bar{\sigma}\,d\dot{\varepsilon}} = \frac{\dot{\varepsilon}\,\bar{\sigma}\,d\,ln\bar{\sigma}}{\bar{\sigma}\,\dot{\varepsilon}\,d\,ln\dot{\varepsilon}} \approx \frac{\Delta\,log\bar{\sigma}}{\Delta\,log\dot{\varepsilon}} = m \qquad (1)$$

In Equation (1), $\dot{\varepsilon}$ is the effective strain rate, and $\bar{\sigma}$ is the effective stress. Since m can take a maximum value of unity (linear dissipater), the dissipation capability of a non-linear dissipater through microstructural changes may be expressed in terms of efficiency, given by:

$$\eta = 2m/(m+1) \qquad (2)$$

The three-dimensional variation of η with temperature and strain rate at a steady-state strain creates a processing map, which is generally presented as a constant-efficiency contour map in a temperature–strain rate frame. The map essentially depicts domains where particular microstructural mechanisms occur and regimes where flow instability occurs as per the continuum criterion, given by [20]:

$$\xi(\dot{\varepsilon}) = \frac{\partial\,ln\,[m/(m+1)]}{\partial\,ln\,\dot{\varepsilon}} + m \leq 0 \qquad (3)$$

The domains corresponding to microstructurally "safe" mechanisms like dynamic recrystallization (DRX) and superplasticity are preferred for hot working, the former being the one chosen for bulk hot working, since it causes large-scale microstructural reconstitution. The temperature and strain rate combination for the peak efficiency in DRX domain may be chosen for optimum hot workability, and the regimes of flow instability may be avoided. It may be emphasized that the processing maps are sensitive to the initial conditions like chemistry, processing history, and microstructure. For developing processing maps, accurate experimental data on flow stress at different temperatures and strain rates and strains are required, which may be obtained in uniaxial compression for example.

2. Experimental Procedure

In the first step, an Mg–4wt %Al–2wt %Ba–2wt %Ca alloy was prepared with pure elemental metals, by melting them under a protective cover of an Ar + 3% SF_6 gas mixture. When the melt reached a temperature of 720 °C, it was poured into a pre-heated permanent mold and allowed to solidify. The cast billets were 104 mm in diameter and 350 mm in length. The cast billet was cut into slices of about 20 mm thickness, from which cylinders with 10 mm diameter and 15 mm height were machined for uniaxial compression testing. The cylindrical specimens were compressed using a computer-controlled servo-hydraulic test machine (M1000/RK; Dartec Ltd., Bournemouth, UK), at temperatures in the range of 260–500 °C and strain rates in the range of 0.0003–10 s^{-1}, and a text matrix involving seven temperatures and six strain rates. Details of the experimental set-up and test procedure have been provided in an earlier publication [21]. In each test, the compressive deformation was stopped when the true strain reached a value of about 1.0 and the deformed specimen was quenched in water. The adiabatic temperature rise that occurred during the test was measured using a K-type thermocouple embedded in the specimen that was connected to the controller of the test machine, and the flow stress values were corrected using flow stress variation with a measured temperature at the selected strain levels [18]. The flow stress values at different temperatures and strain rates at a given strain are used to develop the processing map using the procedure described in an earlier publication [18]. The microstructure of the as-cast material and the deformed specimens was recorded by following standard procedures of polishing and by etching with an aqueous solution containing 3 g picric acid, 20 mL acetic acid, 50 mL ethanol, and 20 mL distilled water.

In the second step, the cast ABaX422 alloy billet with 104 mm diameter was subjected to indirect extrusion in a horizontal hydraulic press, which had a container with a diameter of 110 mm pre-heated to a temperature of 400 °C. The die containing the cast billet was heated to a temperature of 400 °C in an external resistance furnace, and the billet was preheated to 390 °C. Rods of 12 mm diameter (extrusion ratio of 84) were extruded using a ram speed of 1 mm s^{-1}. The extrusion process was designed on the basis of the results from the processing map developed for the as-cast alloy in step 1, as discussed in the following sections. From the extruded rod, cylindrical specimens were prepared for compression testing, and processing maps were developed using the procedure described earlier [18]. For measuring the compressive strength property of the alloy in the as-cast and extruded conditions, compressive tests were conducted in the temperature range 25–250 °C and at a strain rate of 0.0003 s^{-1}. Tensile tests were also conducted on cylindrical specimens of 6 mm diameter with a gage length of 36 mm under select temperature and strain rate conditions, for the purpose of confirming the mechanisms in various workability windows.

3. Results and Discussion

3.1. First Step: Hot Working of Cast ABaX422

Since the hot working response of the alloy depends on the initial microstructure of the material, it is characterized in detail. The microstructure of the as-cast ABaX422 alloy is shown in Figure 1.

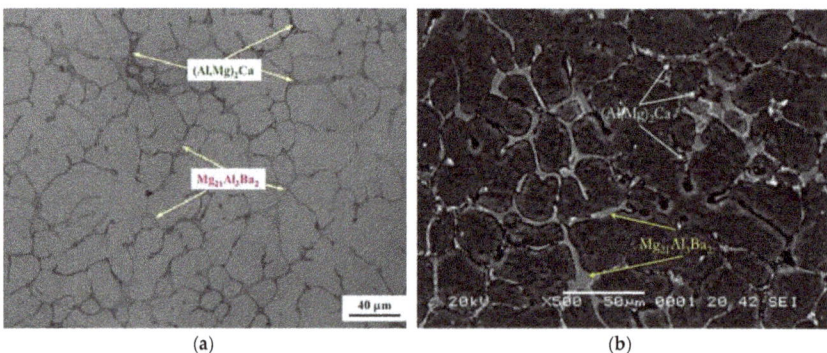

(a)   (b)

Figure 1. (**a**) Optical microstructure and (**b**) SEM micrograph reveals the morphologies of the second phases in the ABaX422 alloy in the as-cast condition.

The average grain diameter is about 25 μm. The image from the scanning electron microscope (SEM) (JOEL 5600, JOEL Ltd, Akishima, Japan), shown in Figure 1b, reveals two types of second phases. These intermetallic phases are mostly present in the grain boundaries as continuous networks. The lamellar phase is enriched with Al and Ca, and identified as $(Mg,Al)_2Ca$, whereas the white blocky phase is $Mg_{21}Al_3Ba_2$ [10].

The true stress–true strain curves obtained at 380 °C and 460 °C on as-cast ABaX422 specimens are shown in Figure 2. In general, the curves exhibited flow softening, which is greater at higher strain rates and lower temperatures. For experiments conducted at temperatures lower than 300 °C and at higher strain rates, it was found that the curves exhibited large multiple drops in flow stress, indicating the occurrence of intense shear fractures. At higher temperatures and lower strain rates, the curves exhibited a near steady-state flow.

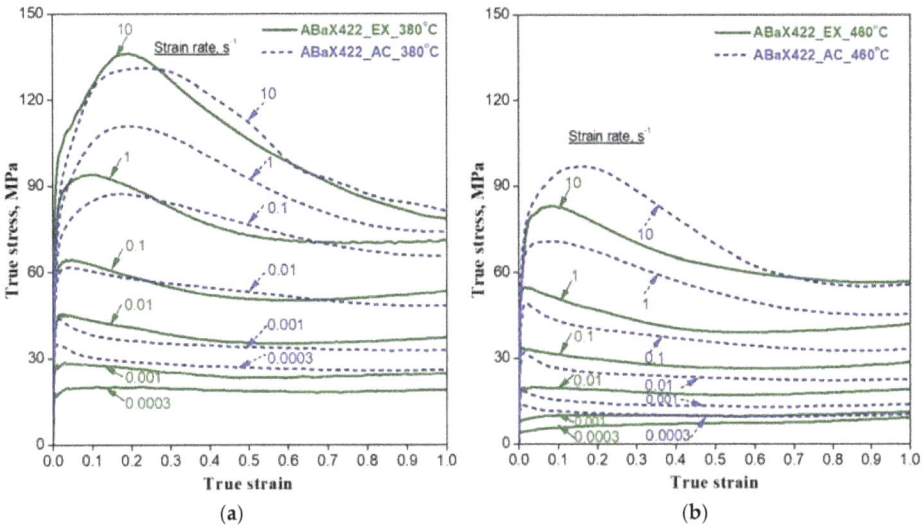

Figure 2. True stress–true strain curves obtained in compression at (a) 380 °C and (b) 460 °C at different strain rates for the ABaX422 alloy, in both as-cast condition and extruded conditions.

The microstructural response (mechanisms) to hot working of as-cast ABaX422 has been characterized using processing map, which is shown in Figure 3. The numbers associated with the contours represent efficiency of power dissipation (Equation (2)) in percent.

The map exhibits two domains in the temperature and strain rate ranges as follows: (1) 300 °C to 390 °C and 0.0003 s^{-1} to 0.001 s^{-1}, with a peak efficiency of 36% occurring at 340 °C and 0.0003 s^{-1}; and (2) 400 °C to 500 °C and 0.0003 s^{-1} to 0.3 s^{-1}, with a peak efficiency of 41% occurring at 500 °C and 0.0003 s^{-1}.

On the basis of microstructures recorded on the deformed specimens and detailed kinetic analysis of the temperature and strain rate dependence of flow stress, these two domains have been identified to represent DRX [13,14]. In the first domain, DRX is controlled by the climb of edge dislocations occurring by lattice self-diffusion, while in the second domain, DRX is controlled by the cross-slip of screw dislocations. Although hot working operations may be conducted in either of the domains, the first domain is not suitable, since the strain rates at which it occurs are too low for the manufacturing process to be viable, and if pushed to higher strain rates, the material undergoes unstable flow resulting in defective microstructure. On the other hand, the second domain is wider in terms of temperature and strain rate ranges; the higher temperatures give better workability and the higher strain rates make the operations faster. To improve the workability of the as-cast material and increase the speed of the manufacturing of wrought products, the constitutive response may be changed by carefully converting the microstructure by hot working at the right temperature and strain rate in Domain #2. The preferred process to achieve this for low-workability materials is the extrusion process, since it is done under constrained compression in a container with large strains imposed in a unit operation and with good control of temperature and speed. The design of such an extrusion process is described below.

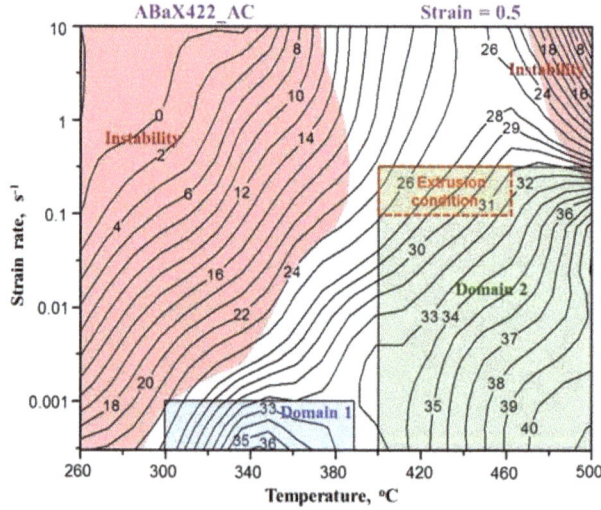

Figure 3. Processing map for the as-cast Mg–4Al–2Ba–2Ca alloy at a true strain of 0.5. The numbers associated with the contours indicate efficiency of power dissipation in percent, and the instability area is shaded in red. The dotted orange color box indicates the extrusion process condition utilized.

3.2. Step 2: Design of Extrusion Experiment

It may be mentioned that the high extrusion ratio is preferred for breaking the as-cast structure completely, and the indirect extrusion process reduces the frictional effects to some extent. The two important process variables in extrusion are the ram speed and the billet temperature. For a simple round-to-round extrusion, the mean strain rate is given by [22]:

$$\dot{\varepsilon} = \frac{\bar{\varepsilon}}{t} = \frac{6v \ln R}{D_b} \tag{4}$$

where $\bar{\varepsilon}$ indicates mean strain (lnR), R is the extrusion ratio, t is time, v represents ram velocity and D_b is the billet diameter. To extrude a bar of 12 mm diameter, the extrusion ratio is 1:84, and for a ram velocity of 1 mm s^{-1} and a billet diameter of 110 mm, the average strain rate is about 0.25 s^{-1}, which is close to the upper end of the strain rate for Domain #2. The local strain rate from surface to center of the billet may vary within a narrow band. Thus, a ram velocity of 1 mm s^{-1} is selected for the extrusion.

As regards the selection of extrusion temperature, factors like an increase in temperature, due to deformation heat and friction at the die wall, and temperature loss, due to conduction to the tools, will have to be considered. For frictionless deformation, the temperature of deformation T_d may be calculated using the equation [22]:

$$T_d = \frac{\bar{\sigma} \, \bar{\varepsilon} \, \beta}{\rho c} \tag{5}$$

where $\bar{\sigma}$ is the mean flow stress, $\bar{\varepsilon}$ is the mean strain (lnR), β is the conversion factor (fraction of deformation converted into heat), ρ represents density, and c represents the specific heat of the work piece. At 400 °C and 0.1 s^{-1}, the flow stress is about 70 MPa, the mean strain for an extrusion ratio of 84 is 4.43, and the density of Mg is 1700 kg m^{-3}, c is equal to 1020 J kg^{-1} K^{-1}, the conversion factor is about 0.85 for hot working, and the calculated temperature increase is about 150 °C. However, due to slower extrusion speed, about half of this may be conducted away to the tools, keeping the temperature increase to about 75 °C. With the starting billet temperature of about 390 °C, the extrusion temperature will rise to a maximum of about 465 °C. Thus, the temperature range of 400–465 °C for extrusion will keep the process within the limits for Domain #2. The corresponding extrusion window is marked on the processing map given in Figure 3.

3.3. Hot Working Behavior of ABaX422 after Extrusion

The microstructure of the extruded alloy is shown in Figure 4. When compared with the as-cast microstructure (Figure 2), the intermetallic phases got refined and redistributed. The grain size is reduced to about 7.5 μm from 25 μm in the as-cast condition. The overall refinement and reconstitution of microstructure caused by the occurrence of DRX during extrusion has a significant influence on workability in the subsequent step, as discussed below.

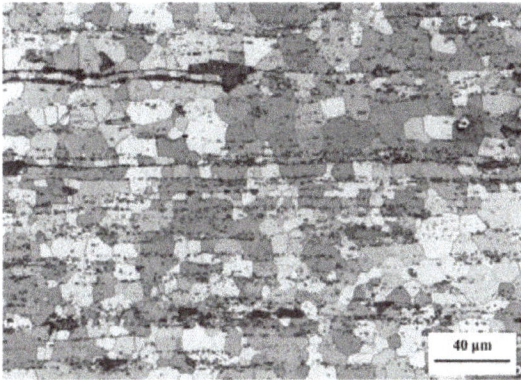

Figure 4. Optical microstructure of the ABaX422 alloy in the extruded condition. The extrusion direction is horizontal.

The stress–strain curves obtained at 380 °C and 460 °C and at different strain rates on the extruded ABaX422 alloy is shown in Figure 2, along with the data on the as-cast alloy. The shapes of the curves are very similar for both conditions, although the strength values are generally higher for the as-cast material, particularly at the lower strain rates. The processing map generated for the extruded ABaX422 alloy that corresponds to a true strain of 0.5 is shown in Figure 5.

The numbers on the contours indicate the efficiency of power dissipation in percent. The reddish color area in the Figure 5 corresponds to the regime of flow instability. The processing map exhibits four domains, in the following temperature and strain rate ranges:

Domain 1: 300–400 °C and 0.0003–0.003 s^{-1}, with a peak efficiency of 32% at 380 °C/0.0003 s^{-1};
Domain 2: 410–500 °C and 0.0003–0.05 s^{-1}, with a peak efficiency of 52% at 500 °C/0.0003 s^{-1};
Domain 3: 360–420 °C and 0.2–10 s^{-1}, with a peak efficiency of 35% at 380 °C/10 s^{-1};
Domain 4: 440–500 °C and 0.2–10 s^{-1}, with a peak efficiency of 40% at 500 °C/10 s^{-1}.

The microstructures obtained for specimens deformed at peak efficiency conditions in the four domains of the processing map are shown in Figure 6. All of them show that DRX has occurred in all the domains. The grain size in the lower temperature domains 1 and 3 is finer than that in the higher temperature domains 2 and 4. The tensile ductility (total elongation) values measured near the peak efficiency conditions are 47% in Domain 1, highest (84%) in Domain 2, 35% in Domain 3, and 48% in Domain 4. Tensile flow curves are shown in Figure 7, and the corresponding mechanical properties are listed in Table 1. The fractographs obtained on the fracture surfaces of the tensile specimens are shown in Figure 8, which indicate that ductile fracture has occurred in Domains 1, 2, and 3, while features of the intercrystalline fracture are seen in Domain 4. Dimple features that can be seen in Figure 8a–c typically indicate ductile fractures.

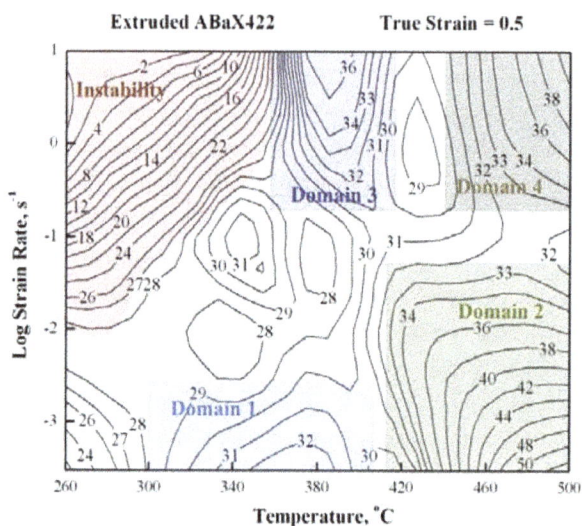

Figure 5. Processing map for the extruded ABaX422 alloy developed at a true strain of 0.5. The numbers shown with the contours represent dissipation efficiency in percent. The shaded reddish area represents the flow instability regime.

Table 1. The tensile mechanical properties of the specimens that correspond to the four domains in the processing maps of the extruded ABaX422 alloy.

Condition	Yield Strength (MPa)	Ultimate Tensile Strength (UTS), MPa	Strain to Fracture	Maximum Displacement (mm)	% Elongation
Domain 1 (380 °C and 0.0003 s^{-1})	41	50	0.47	19.25	47%
Domain 2 (500 °C and 0.0003 s^{-1})	2	6	0.82	32.89	84%
Domain 3 (380 °C/2.3 s^{-1})	80	121	0.35	17.61	35%
Domain 4 (500 °C/2.3 s^{-1})	50	64	0.49	19.64	48%

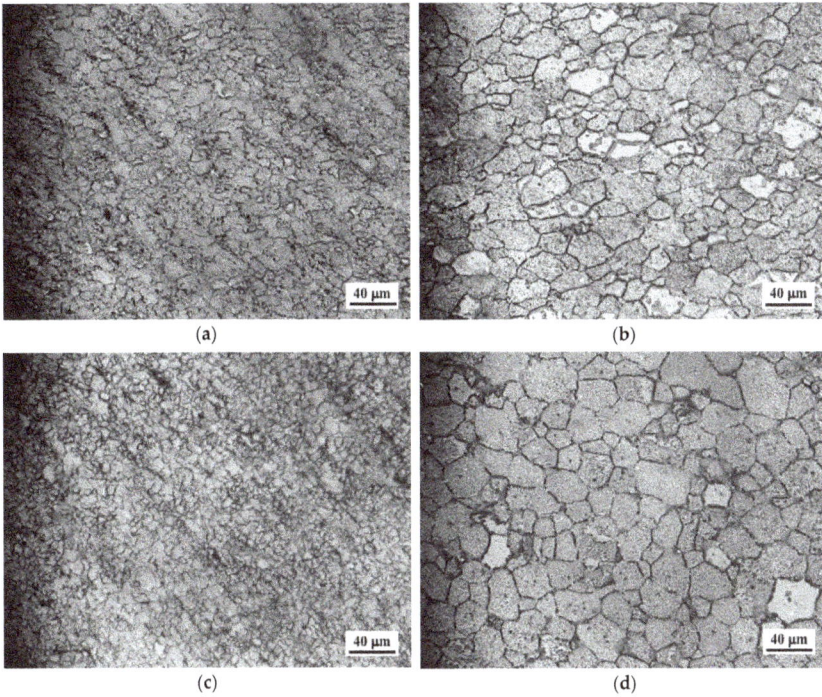

Figure 6. Microstructures obtained on extruded ABaX422 alloy specimens compressed under peak efficiency conditions in the four different domains of the processing map. (**a**) Domain 1 (380 °C/0.0003 s^{-1}), (**b**) Domain 2 (500 °C/0.0003 s^{-1}), (**c**) Domain 3 (380 °C/10 s^{-1}), and (**d**) Domain 4 (500 °C/10 s^{-1}).

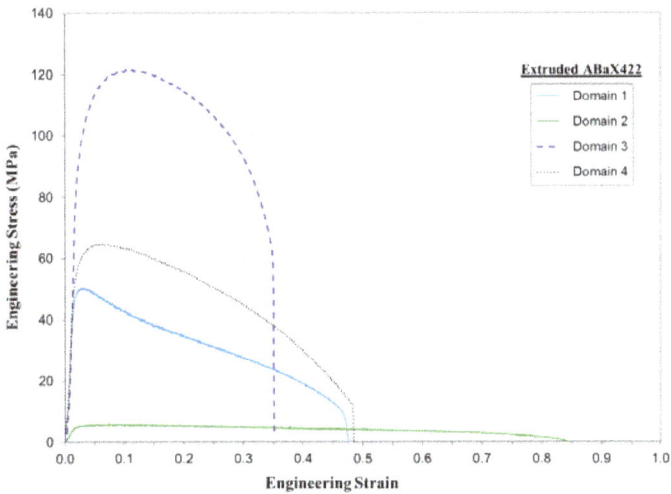

Figure 7. Tensile flow curves for the specimens deformed in the four domains of the processing maps: Domain 1 (380 °C/0.0003 s^{-1}), Domain 2 (500 °C/0.0003 s^{-1}), Domain 3 (380 °C/2.3 s^{-1}), and Domain 4 (500 °C/2.3 s^{-1}).

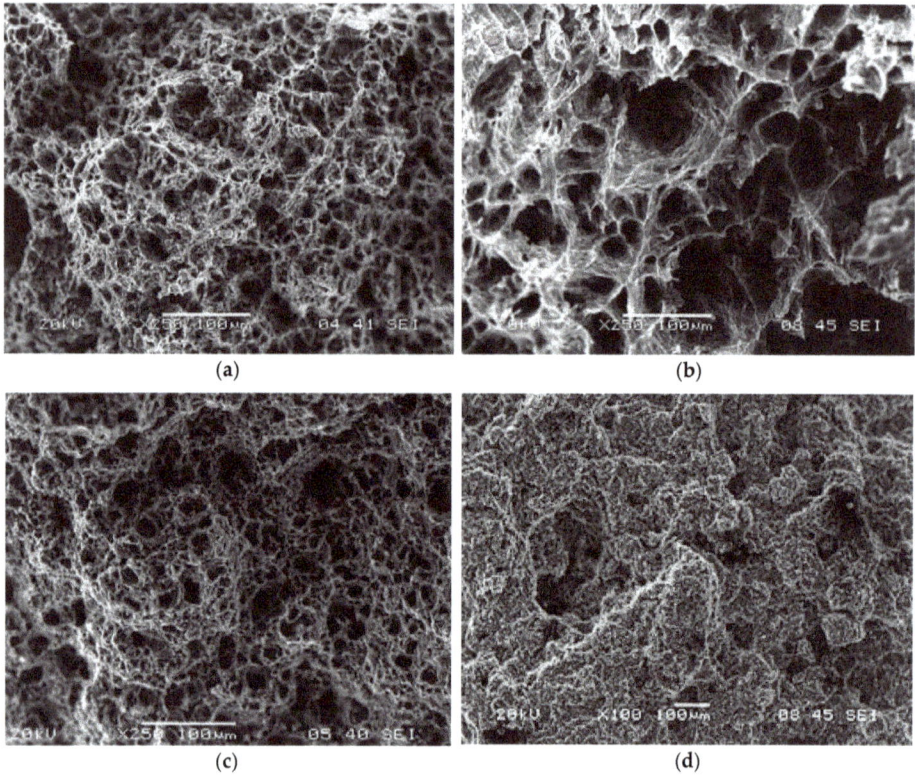

Figure 8. SEM fractographs of tensile specimens deformed in the four domains of the processing maps: (**a**) Domain 1 (380 °C/0.0003 s^{-1}; total elongation: 47%), (**b**) Domain 2 (500 °C/0.0003 s^{-1}, total elongation: 84%, (**c**) Domain 3 (380 °C/2.3 s^{-1}; Total elongation: 35%), and (**d**) Domain 4 (total elongation: 48%).

The mechanisms of DRX in Domains 1, 2, and 3 may be further analyzed with the help of kinetic analysis, using a kinetic rate equation that relates the steady-state flow stress (σ) to the temperature (T) and strain rate ($\dot{\varepsilon}$), given by [23]:

$$\dot{\varepsilon} = A\sigma^n \exp[-Q/RT] \tag{6}$$

where n, Q, and R are the stress exponent, apparent activation energy, and gas constant, respectively, and A is a constant. Since the rate equation is obeyed within the deterministic domains, the apparent activation energy value can be evaluated for each domain. A plot of flow stress versus strain rate at different temperatures is shown in Figure 9a, and the Arrhenius plot showing the natural logarithm of flow stress normalized with respect to the shear modulus versus the inverse of the absolute temperature is shown in Figure 9b. The values of the activation parameters obtained from these plots are shown in Table 2, for extruded as well as as-cast conditions.

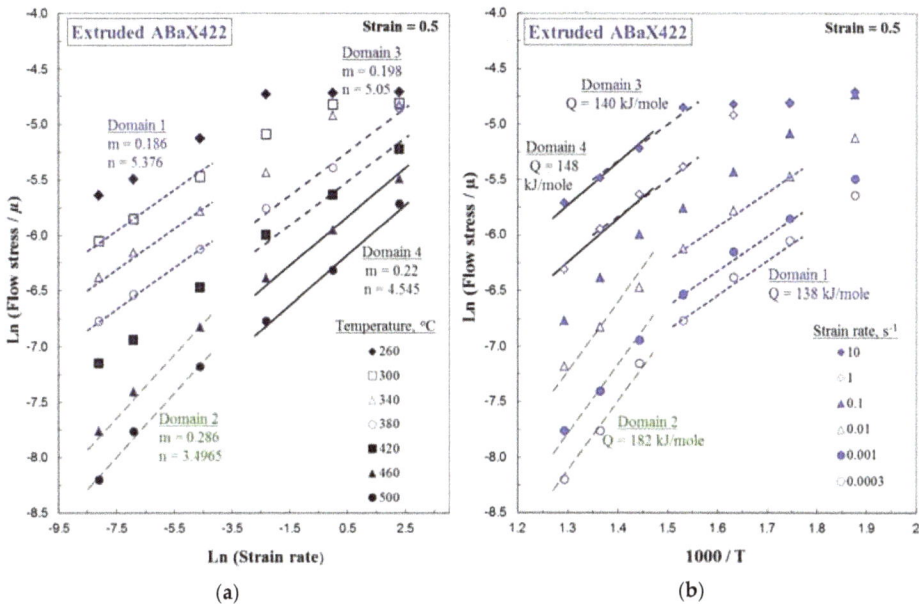

Figure 9. (a) Variation of normalized flow stress with strain rate on natural logarithmic scale, and (b) an Arrhenius plot of normalized flow stress with the inverse of absolute temperature. The apparent activation parameters n and Q are estimated with data relevant to various domains of the processing map for the extruded ABaX422 alloy.

Table 2. The kinetic parameters, namely the stress exponent (n) and apparent activation energy (Q), in the various domains of the processing maps for the ABaX422 alloy, and the proposed rate-controlling mechanisms in the as-cast and extruded conditions.

Domain #	As-Cast (Ref. [13])			Extruded		
	n	Q (kJ/mole)	Mechanism	n	Q (kJ/mole)	Mechanism
1	5.24	169	Climb (LSD)	5.37	138	Climb (LSD)
2	4.46	263	Cross-slip	3.49	182	Cross-slip
3	-	-	-	5.05	140	Climb (GSD)
4	-	-	-	4.54	148	GB cracking

Note: LSD: Lattice self-diffusion; GSD: Grain boundary self-diffusion; GB: grain boundary.

In Domain 1, which occurs in the temperature range of 300–400 °C, a basal slip on {0001} <11$\bar{2}$0> and a prismatic slip {10$\bar{1}$0} <11$\bar{2}$0> occur, and the recovery mechanism that nucleates DRX is climb of edge dislocations, which is controlled by the lattice self-diffusion (LSD). The estimated apparent activation energy (138 kJ/mole) is close to that of lattice self-diffusion in Mg, which is 135 kJ/mole [24]. The slower strain rates at which this domain occurs favors lattice self-diffusion. Thus, DRX in Domain 1 occurs by basal and prismatic slip, along with climb mechanism by lattice self-diffusion.

In Domain 2, which occurs at higher temperatures (410–500 °C), a second-order pyramidal slip {11$\bar{2}$2} <11$\bar{2}$3> occurs. The recovery mechanism nucleating DRX for this system is cross-slip, since a large number of intersecting slip systems are available and the stacking fault energy on these slip planes is considerably high (173 mJ m^{-2}) [25]. The activation energy for this process is higher than that for self-diffusion, due to the particles present in the matrix causing back stress, and the value estimated in this domain (182 kJ/mole) is in support of this. Thus, the mechanism of DRX in Domain 2 involves a second-order pyramidal slip, along with a cross-slip for recovery.

In Domain 3, which occurs at lower temperatures similar to Domain 1, a basal slip {0001} <11$\bar{2}$0> and a prismatic slip {10$\bar{1}$0} <11$\bar{2}$0> occur. The recovery process is a climb of the edge dislocation for these slip systems. However, since this domain occurs at higher strain rates, lattice self-diffusion cannot be a rate-controlling process for the climb, since it is too slow of a process. Alternately, grain boundary self-diffusion may be expected to occur, since it is faster and favored by the finer grain size in the extruded alloy. The apparent activation energy is 140 kJ/mole, which is higher than that required for grain boundary self-diffusion (95 kJ/mole) [24]. The large amount of particle content present in the microstructure causes a high back stress, which increases the apparent activation energy. Thus, the mechanism of the DRX process in Domain 3 is basal and prismatic slip, along with climb controlled by grain boundary self-diffusion.

In Domain 4, the intercrystalline fracture features (Figure 8d) suggest that this domain represents intergranular fracture. Since this domain occurs at higher temperatures, a second-order pyramidal slip {11$\bar{2}$2} <11$\bar{2}$3> occurs. However, the stress concentration at the grain boundaries is not relieved by recovery, since the occurrence of cross-slip is restricted due to high strain rates, resulting in intergranular fracture.

3.4. Hot Working Behavior of As-Cast Versus Extruded ABaX422 Alloy

The differences in the constitutional response of the ABaX422 alloy in as-cast and extruded conditions may be analyzed by comparing the respective processing maps given in Figures 3 and 5. Firstly, Domains 1 and 2 have essentially similar characteristics, including the kinetic parameters given in Table 2. The apparent activation energy is higher for the as-cast alloy, which may be attributed to the back stress generated by the coarse intermetallic phases at the grain boundaries. Secondly, two additional domains have appeared in the map for the extruded alloy. Domain 4 has been identified to represent intercrystalline fracture, and hence is not useful for hot working. However, Domain 3 is worth considering for manufacturing purposes, because it covers faster strain rates (up to 10 s^{-1}); therefore, manufacturing is viable and the product will have a fine-grained microstructure, which is highly desirable.

A careful examination of the two processing maps also reveals that the flow instability regimes are reduced by extruding the material, which gives flexibility in terms of widening the hot working parameters without the onset of instabilities like adiabatic shear banding or flow localization in the product. The presence of these instability manifestations will result in inferior and inconsistent mechanical properties, and should be avoided. Thus, by introducing a hot extrusion step after casting, not only may the material's workability be improved but also better components may be manufactured faster, making the extrusion a viable process.

3.5. Connected Process Design

An ABaX422 alloy is used for applications in the automobile industry where creep resistance is a critical requirement. For applications like engine blocks, casting is the primary process. However, for applications where structural integrity and large-scale manufacturing is required, further mechanical working of an as-cast alloy is restricted by its low hot workability and lower speeds of processing. The above results show that by designing a connected step of hot extrusion under controlled conditions, the constitutive response of the alloy may be changed to enhance its workability, mechanical properties, and speed of processing, such that further manufacturing becomes viable. It may be mentioned that a high extrusion ratio is preferred for breaking the as-cast structure completely, and the indirect extrusion process reduces the frictional effects to some extent. The use of high extrusion ratios may result in a smaller diameter or small-size product, but the process may be designed by suitable modification of extrusion equipment, in order to produce the desired sizes. The microstructural changes that are responsible for the change in the material's response to hot working through this connected process design are summarized in Figure 10, which reveals how the cast structure gets refined in these steps, resulting in better properties in the product. Once the cast structure is transformed into a wrought structure by extrusion (step 2), further hot working by upset forging—for example, at 380 °C and 10 s^{-1} (step 3)—results in a highly desirable fine-grained microstructure.

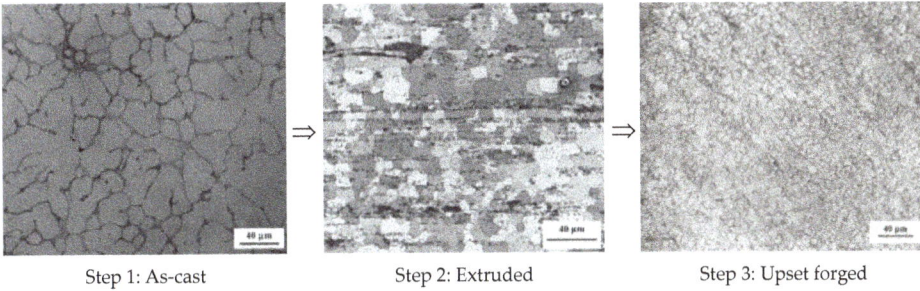

| Step 1: As-cast | Step 2: Extruded | Step 3: Upset forged |

Figure 10. Microstructural changes in continued processing steps of the ABaX422 magnesium alloy.

3.6. Compressive Strength of ABaX422 in Temperature Range 25–250 °C

The compressive yield and ultimate strength of the extruded alloy are given in Figure 11 and compared with that for the as-cast alloy. By extruding the alloy, its strength was higher than that of the as-cast alloy up to a temperature of about 175 °C, and the ultimate compressive strength exhibited an improvement of about 42% at room temperature. The enhanced strength property may be attributed to the grain refinement and redistribution of the intermetallic particles caused by the extrusion shown in Figure 4. The generation of crystallographic texture is an additional factor that influences the strength changes, since the as-cast alloy is nearly random, while the extruded rod has a preferred orientation.

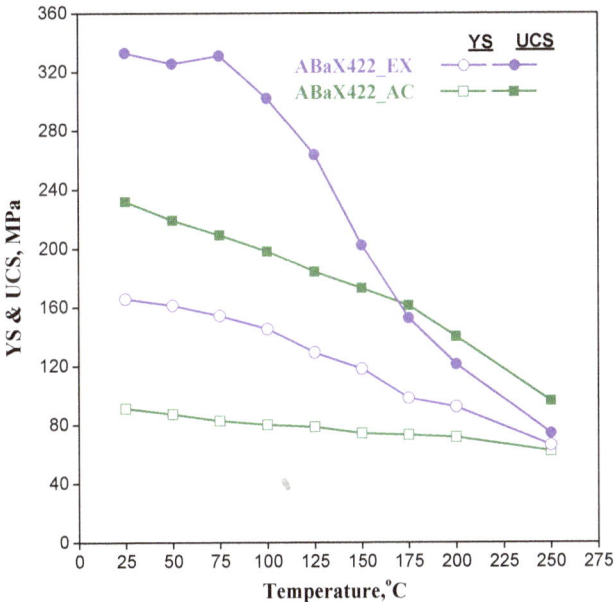

Figure 11. Compressive yield and ultimate compressive strength of the ABaX422 alloy in the as-cast and extruded conditions.

4. Conclusions

The constitutive response of the ABaX422 alloy to hot working has been characterized for both as-cast and extruded conditions with the help of processing maps, with a view to design connected processing steps that ensure the viable manufacturing of components. The following conclusions are drawn from this investigation:

1. ABaX422 alloy in the as-cast condition has a limited workability, due to a coarse and large-grained microstructure, and cannot be hot worked at higher speeds without causing microstructural damage.
2. The processing map for the as-cast ABaX422 alloy offers a window, at temperatures higher than 400 °C and strain rates lower than 0.3 s^{-1}, where the alloy may be hot worked.
3. A connected step of extrusion has been designed by selecting a process parameter as per the processing map for the as-cast alloy, and extrusion of a 104 mm diameter billet has been conducted to produce a 12 mm diameter rod product, the microstructure of which has a finer grain size with a redistributed fine particles of the intermetallic phases.
4. The processing map for the extruded alloy exhibited four domains—two of them, representing dynamic recrystallization, are similar to those exhibited in the processing map for as-cast material, while the remaining two only appear with the extruded material.
5. Out of the two new domains exhibited in the map for the extruded alloy, the one occurring in the temperature range 360–420 °C and strain rate range 0.2–10 s^{-1} (Domain 3) is useful for manufacturing, since the strain rate is higher, making the process viable, and the temperature is lower, resulting in a finer grain size in the product.
6. The fourth domain that occurs at temperatures higher than 440 °C and strain rates higher than 0.2 s^{-1} represents intercrystalline cracking and causes reduced workability.
7. The area of the flow instability regime exhibited in the processing map for the as-cast alloy is reduced by the extrusion step, and this enlarges the workability window.
8. Connected process design by changing the constitutive response of hard-to-process alloys may be used to manufacture wrought components at viable speeds and with better mechanical properties.

Author Contributions: K.P.R. and C.D. performed the analysis of the data, generating the processing maps and kinetic analysis, and writing the paper; K.S. performed the experimental work, generating the results and microstructural work; Y.V.R.K.P. contributed to the aspects related to the processing maps and writing the paper; H.D. and N.H. developed and provided the alloys in their initial cast and extruded forms, as well as their microstructures.

Acknowledgments: This work described in this paper was supported by a grant from the Strategic Research Grant (Project #7002744) from the City University of Hong Kong.

Conflicts of Interest: The authors declare no conflict of interest.

References

1. Pekguleryuz, M.; Celikin, M. Creep resistance in magnesium alloys. *Int. Mater. Rev.* **2010**, *55*, 197–217. [CrossRef]
2. Powell, B.R.; Rezhets, V.; Balogh, M.P.; Waldo, R.A. Microstructure and creep behavior in AE42 magnesium die-casting alloy. *JOM* **2002**, *54*, 34–38. [CrossRef]
3. Amberger, D.; Eisenlohr, P.; Göken, M. Influence of microstructure on creep strength of MRI 230D Mg alloy. *J. Phys. Conf. Ser.* **2010**, *240*, 012068. [CrossRef]
4. Suzuki, A.; Saddock, N.D.; Riester, L.; Lara-Curzio, E.; Jones, J.W.; Pollock, T.M. Effect of Sr additions on the microstructure and strength of a Mg-Al-Ca ternary alloy. *Metall. Mater. Trans. A* **2007**, *38A*, 420–427. [CrossRef]
5. Rzychon, T.; Chmiela, B. The influence of tin on the microstructure and creep properties of a Mg-5Al-3Ca-0.7Sr-0.2Mn magnesium alloy. *Solid State Phenom.* **2012**, *191*, 151–158. [CrossRef]

6. Hirai, K.; Somekawa, H.; Takigawa, Y.; Higashi, K. Effect of Ca and Sr addition on mechanical properties of a cast AZ91 magnesium alloy at room temperature and elevated temperature. *Mater. Sci. Eng. A* **2005**, *403*, 276–280. [CrossRef]

7. Sato, T.; Kral, M.V. Microstructural evolution of Mg-Al-Ca-Sr alloy during creep. *Mater. Sci. Eng. A* **2008**, *498*, 369–376. [CrossRef]

8. Sadeghi, A.; Pekguleryuz, M. Recrystallization and texture evolution of Mg-3%Al-1%Zn-(0.4-0.8%)Sr alloys. *Mater. Sci. Eng. A* **2011**, *528*, 1678–1685. [CrossRef]

9. Dieringa, H.; Hort, N.; Kainer, K.U. Barium as alloying element for a creep resistant magnesium alloy. In Proceedings of the 8th International Conference on Magnesium Alloys and their Applications, Weinheim, Germany, 26–29 October 2009; Kainer, K.U., Ed.; pp. 62–67.

10. Dieringa, H.; Huang, Y.; Wittke, P.; Klein, M.; Walther, F.; Dikovits, M.; Poletti, C. Compression-creep response of magnesium alloy DieMag422 containing barium compared with the commercial creep-resistant alloys AE42 and MRI230D. *Mater. Sci. Eng. A* **2013**, *585*, 430–438. [CrossRef]

11. Dieringa, H.; Zander, D.; Gibson, M.A. Creep behaviour under compressive stresses of calcium and barium containing Mg-Al-based die casting alloys. *Mater. Sci. Forum* **2013**, *765*, 69–73. [CrossRef]

12. Rao, K.P.; Suresh, K.; Prasad, Y.V.R.K.; Hort, N.; Dieringa, H. Microstructural Response to Hot Working of Mg-4Al-2Ba-1Ca (ABaX421) as Revealed by Processing Map. In Proceedings of the 10th International Conference on Magnesium Alloys and Their Applications, Jeju, Korea, 11–16 October 2015; pp. 97–104.

13. Rao, K.P.; Ip, H.Y.; Suresh, K.; Prasad, Y.V.R.K.; Wu, C.M.L.; Hort, N.; Kainer, K.U. Compressive strength and hot deformation mechanisms in as cast Mg-4Al-2Ba-2Ca (ABaX422) alloy. *Philos. Mag.* **2013**, *93*, 4364–4377. [CrossRef]

14. Suresh, K.; Rao, K.P.; Prasad, Y.V.R.K.; Wu, C.-M.L.; Hort, N.; Dieringa, H. Mechanism of dynamic recrystallization and evolution of texture in the hot working domains of the processing map for Mg-4Al-2Ba-2Ca alloy. *Metals* **2017**, *7*, 539. [CrossRef]

15. Rao, K.P.; Dharmendra, C.; Prasad, Y.V.R.K.; Hort, N.; Dieringa, H. Optimization of thermo-mechanical processing for forging of newly developed creep-resistant magnesium alloy ABaX633. *Metals* **2017**, *7*, 513. [CrossRef]

16. Rao, K.P.; Lam, S.W.; Hort, N.; Dieringa, H. High Temperature Deformation Behavior of a Newly Developed Mg Alloy Containing Al, Ba and Ca. In Proceedings of the 7th Thai Society of Mechanical Engineers, International Conference on Mechanical Engineering, Chiang Mai, Thailand, 13–16 December 2016; p. 6, AMM0023.

17. Prasad, Y.V.R.K.; Seshacharyulu, T. Modelling of hot deformation for microstructural control. *Int. Mater. Rev.* **1998**, *43*, 243–258. [CrossRef]

18. Prasad, Y.V.R.K.; Rao, K.P.; Sasidhara, S. *Hot Working Guide: A Compendium of Processing Maps*, 2nd ed.; ASM International: Materials Park, OH, USA, 2015; ISBN 978-1-62708-091-0.

19. Prasad, Y.V.R.K. Processing maps: A status report. *J. Mater. Eng. Perform.* **2003**, *12*, 638–645. [CrossRef]

20. Ziegler, H. *Progress in Solid Mechanics*; Sneddon, I.N., Hill, R., Eds.; John Wiley: New York, NY, USA, 1965; Volume 4, pp. 91–193.

21. Prasad, Y.V.R.K.; Rao, K.P. Processing maps and rate controlling mechanisms of hot deformation of electrolytic tough pitch copper in the temperature range 300–950 °C. *Mater. Sci. Eng. A* **2005**, *391*, 141–150. [CrossRef]

22. Dieter, G.E. *Mechanical Metallurgy, SI Metric Edition*; McGraw Hill Book Co.: London, UK, 1988; p. 628 & p. 525.

23. Jonas, J.J.; Sellars, C.M.; Tegart, W.M. Strength and structure under hot working conditions. *Metall. Rev.* **1969**, *14*, 1–24. [CrossRef]

24. Frost, H.J.; Ashby, M.F. *Deformation-Mechanism Maps*; Pergamon Press: Oxford, UK, 1982; p. 44.

25. Morris, J.R.; Scharff, J.; Ho, K.M.; Turner, D.E.; Ye, Y.Y.; Yoo, M.H. Prediction of a {1122} hcp stacking fault using a modified generalized stacking-fault calculation. *Philos. Mag. A* **1997**, *76*, 1065–1077. [CrossRef]

metals

MDPI

Article

Assessment of Metal Flow Balance in Multi-Output Porthole Hot Extrusion of AA6060 Thin-Walled Profile

Xin Xue [1,2], Gabriela Vincze [2], António B. Pereira [2] ⓘ, Jianyi Pan [3] and Juan Liao [1,2,*]

[1] School of Mechanical Engineering and Automation, Fuzhou University, Fuzhou 350116, Fujian, China; xin@fzu.edu.cn
[2] Centre for Mechanical Technology and Automation, Department of Mechanical Engineering, University of Aveiro, 3810-193 Aveiro, Portugal; gvincze@ua.pt (G.V.); abastos@ua.pt (A.B.P.)
[3] School of Mechanical Engineering, Guangzhou College of South China University of Technology, Guangzhou 510800, China; panjianyi@gcu.edu.cn
* Correspondence: jliao@fzu.edu.cn or jliao@ua.pt; Tel.: +86-591-2286-6793

Received: 15 May 2018; Accepted: 10 June 2018; Published: 18 June 2018

Abstract: For the porthole hot extrusion of a thin-walled tube based on metal flow, the role of the die's structure should be focused on to achieve precision formation, especially for multi-output extrusion and/or complex cross-sectional profiles. In order to obtain a better metal flow balance, a multi-output porthole extrusion die was developed, including some novel features such as a circular pattern of the portholes with a dart-shaped inlet bridge, a buckle angle in the inlet side of the upper die, a two-step welding chamber, and a non-uniform bearing length distribution. Through the use of thermo-mechanical modeling combined with the Taguchi method, the underlying effects of key die features were investigated, such as the billet buckle angle, the porthole bevel angle, the depth of the welding chamber, and the type of bridge on the metal flow balance. The experimental validation showed that the developed numerical model for the multi-output porthole extrusion process had high prediction accuracy, and was acceptable for use in an industrial extrusion with a complex section.

Keywords: multi-output porthole extrusion; aluminum alloy; thin-walled profile; metal flow; optimization

1. Introduction

Fuel savings and a reduction in emissions can be achieved by using lightweight materials and the improvement of their formation processes. Aluminum alloys, which have superior mechanical properties such as low weight, excellent corrosion resistance, favorable formability, etc., are widely used in various industries [1]. With a need for a reduction in the weights of parts, porthole die extrusion increased a demand for hollow cross-section profiles characterized by a complex multi-cavity and thinner walls [2–4]. However, porthole die extrusion is a process dominated by metal flow under multiple constraints, including the ram velocity, the temperature field, the friction conditions, the material properties, etc. Satisfactorily extruded profiles can only be achieved under the reasonable coordination of these formation parameters and an appropriate die design. In the past decade, many studies were conducted on the optimization of the porthole extrusion process via theoretical, experimental, and numerical methods [4–6]. Here, the authors focused on the die design in terms of metal flow balance. As a major factor affecting the metal flow, the design of the porthole die structure should play a considerable role in the quality of the final extruded profile.

Multi-output porthole extrusion is one of many manufacturing processes. This formation process reduces the extrusion load and increases the efficiency of mass production, despite the high value of the extrusion ratio (the ratio of input to output cross-sectional areas). Sinha et al. estimated the ram force using the upper bound method while considering the process as a single-hole and a multi-hole extrusion [7,8]. They reported that multi-hole processes are more suitable for the mass production of small-sized components than single-hole solutions. Zhang et al. investigated the effects of the uniformity of the material flow on the extrusion load via the signal-to-noise ratio using the Taguchi method [9]. The effects of several process parameters were examined, including the billet diameter, the ram speed, the temperature fields for the metal flow, and the extrusion force. Sun et al. studied the non-uniform deformation path of materials and the formation of a defect generation mechanism by means of a multi-way loading of multi-cavity parts [10]. They reported that the material flow behavior and defect generations are sensitive to loading modes, die structures, and billet shapes and sizes.

The abovementioned studies provide useful knowledge for the optimization of the formation of the porthole extrusion process in terms of various hollow profiles. However, these researchers paid less attention to multi-output porthole hot extrusion and tended to focus on large hollow profiles with complex cross sections rather than on small and highly thin-walled profiles. Recently, some efforts were made with regard to the design schemes of dies for the analyses of metal flow balance and the quality of extruded profiles. Lee et al. investigated the effects of a variation in chamber shape on the porthole die extrusion process, specifically looking at metal flow, extrusion load, and the tendency of mandrel deflection [11]. Zhang et al. attempted to optimize die configurations through the location of the die orifices for metal flow, the extrusion load, die deflection, etc., before summarizing some design laws for the three-hole extrusion of an AA6063 tube [12]. Gagliardi et al. investigated the process load and welding quality of AA6086 extruded parts through variations in the tooling geometry [13]. They indicated that a better welding performance and improved tooling residual stress could be obtained. Den Bakker et al. studied the effects of various geometries of the weld chamber on the quality of the longitudinal and transverse weld seams and pointed out that the corresponding mechanical properties were greatly affected by the density of the oxide particle population [14,15]. Zhang et al. d the effects of the extrusion die's structure on the length of the transverse weld of an AA7N01 hollow profile [16]. Yu et al. developed a set of modular porthole extrusion dies with varying depths of welding chambers [17]. Their results indicated that metal flow behavior with a shallow welding chamber led to a macro-hole formation in the extruded profile, which was not caused by the solid-state bonding process. The International Conference on Extrusion and Benchmark (ICEB) conference series was established to provide solutions to the extrusion industry on the basis of discussing numerical techniques. For example, Gamberoni et al. studied the processing conditions for the extrusion of an EN-AW-6063 hollow profile (industrial benchmark) through flow-stress analyses using various material models [18]. Selvaggio et al. investigated mandrel deflection, and local temperature and pressure in industrial extrusion dies [19]. Furthermore, they studied the effects of choking and variations in bearing length as well as the process control in terms of press load, die temperatures, die deflection, and final profile lengths [20]. Numerical simulations were proven to be an efficient tool in the analysis of the porthole extrusion process. However, none of the previously mentioned studies presented a multi-output porthole extrusion for a non-axisymmetric highly thin-walled profile (i.e., a wall thickness less than 1.0 mm), which should result in more complicated metal flow behavior.

The presented work aimed to explore the influence of key die structures on metal flow balance in terms of the developed multi-output porthole extrusion process. A typical non-asymmetric thin-walled tube made from an aluminum alloy was addressed, which had an evident case of profile distortion due to unbalanced metal flow. First, an industrial six-output porthole extrusion die was designed and introduced. Second, an effective thermo-mechanical model of a multi-output porthole extrusion process was established, as well as the experimental validation of the numerical model. Finally, the optimization of the die features in the multi-output porthole extrusion process was performed based on the Taguchi method, while the sources of unbalanced deformation in the multi-output porthole extrusion were analyzed and discussed.

2. Design Scheme of Multi-Output Porthole Extrusion Die

2.1. General Description of the Extruded Profile

The aluminum profile in the presented case study is a typical thin-walled tube with the snap-fit channel for industrial application, as shown in Figure 1a. In fact, the studied profile is mainly utilized in the frames of the flat-panel display products. The main geometries of the studied profile include the outer diameter of 12 mm, the whole length of 19.7 mm, the outer wall thickness of only 0.7, and the middle wall thickness of 0.8 mm. Considered the wear regulation of the extrusion die, the wall thickness of the hollow profile generally should be designed within a minus tolerance. For the cantilever strength reason, the middle wall thickness should be designed as a plus tolerance. In this study, due to the high extrusion ratio (up to 96.7), the multi-output porthole extrusion process has been developed to reduce the forming load and improve the product quality. As shown in Figure 1b, six output profiles can be obtained for each extrusion. If four output profiles were adopted, the extrusion ratio would be up to the value of 150. This means that the multi-output extrusion process is too high for the AA6060 subjected to plastic deformation. In the other side, if eight output profiles were used, too many final output profiles make a difficulty at the next process control, namely towing or drawing of the initial extruded profiles. Therefore, the authors think that the design solution of six output profiles should be a reasonable chosen for this studied case. A typical multi-output porthole extrusion system is mainly composed of three parts: an upper die, a lower die, and a ram. The pre-heated aluminum alloy billet undergoes successively splitting, welding, splitting again, and the profile finally is extruded out of the bearing lands of lower die.

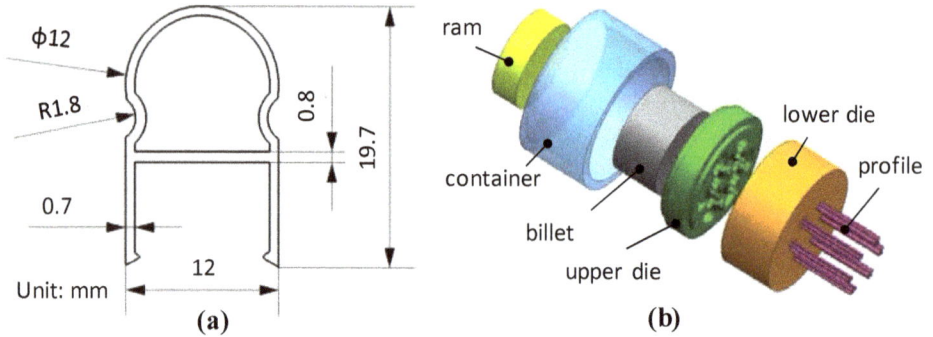

Figure 1. (a) Geometry of thin-walled profile; (b) die configuration for multi-output porthole extrusion.

2.2. The Main Structure Design of the Multi-Output Porthole Extrusion Die

In order to obtain a better material flow system, the circular pattern of the portholes with dart-shaped inlet bridge in the upper die is designed, as shown in Figure 2. The porthole number is 12, while the location of each mandrel is close to the geometric center of three-porthole group. The section shapes of the bridge and the mandrel are shown in Figure 2, as well as their basic geometries. The buckle angle in the inlet side of the upper die is used for billet overflow resistance. Some previous efforts pointed out that these structure designs may help to get a balance metal flow and reduce the forming load during the hot extrusion of aluminum alloys [9,12].

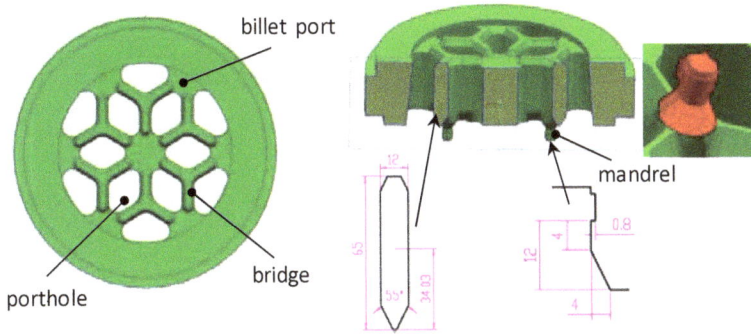

Figure 2. The main structures of upper die.

Figures 3 and 4 show the main structures of the lower die with the two-step welding chamber and the non-uniform bearing length distribution, respectively. Both of design solutions have been proven to be effective in controlling the material flow compared to the previous or traditional design solution, as illustrated in Figure 5. It is reported that the forming load is reduced 10–12%. In this work, some key die features of the multi-output porthole extrusion will be assessed in terms of the material flow and the design scheme.

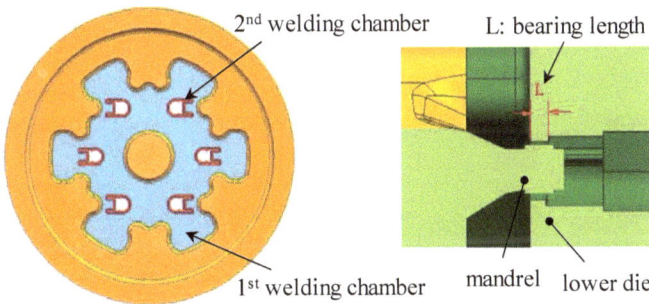

Figure 3. The main structures of lower die.

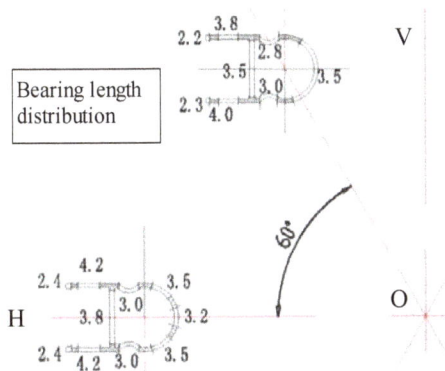

Figure 4. The non-uniform bearing length distribution of lower die.

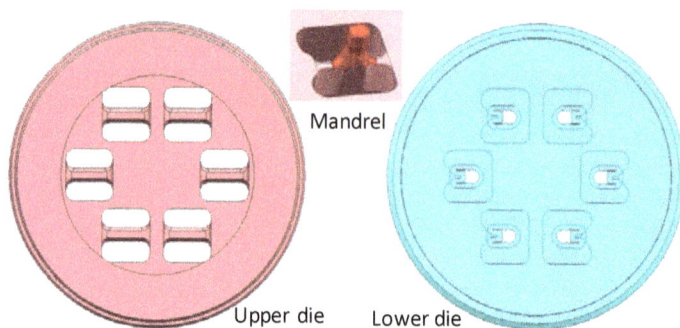

Figure 5. The traditional tool design in terms of the selected multi-porthole extrusion process.

2.3. Design of Experiment by Means of the Taguchi Method

The Taguchi method has been employed widely in various fields in order to optimize die design or process parameters with a significant reduction in cost and time involved [21,22]. Besides the pocket contour in this studied case, there are several die features will affect the extrusion process or the metal flow balance, namely: the billet buckle angle (A), the porthole bevel angle (B), the depth of welding chamber (C), and the bridge type (D), as shown in Figure 6. The selected die feature parameters for the design of experiment (DOE) are given in Table 1. The experimental layout plan with four factors and three levels using L9 orthogonal array was performed to study the effects of the above mentioned die features on the metal flow during the multi-output porthole extrusion including the relative exit velocity of extruded profile and the mandrel deflection.

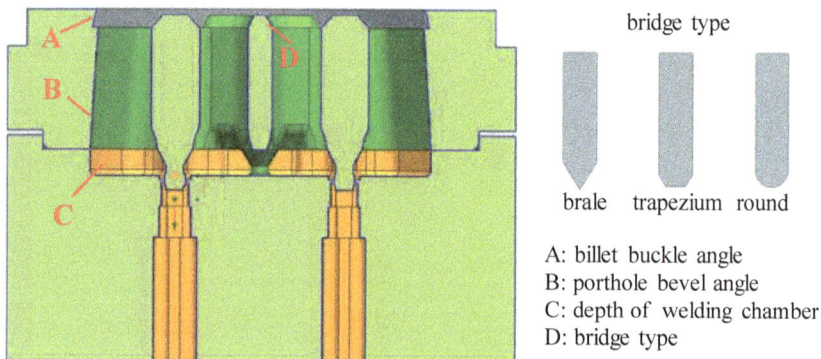

Figure 6. The selected die feature for control factors of Taguchi method.

Table 1. Control factors and levels of die configurations used in the Taguchi method.

	Control Factors	Levels		
		1	2	3
A	Billet buckle angle (α, °)	5	15	30
B	Porthole bevel angle (β, °)	5	7.5	10
C	Depth of welding chamber (d, mm)	10	15	20
D	Bridge type	Brale	Trapezium	Round

The Taguchi method determines the optimum values of each parameter using the SN ratio (Signal-to-Noise ratio). Generally, a larger signal and smaller noise is the best, so that the largest SN ratio yields the optimum result. The SN ratio is calculated according to the characteristics of the output properties, which are "the-smaller-the-better type", "the-larger-the-better type", and "on-target type". Since the output properties of this research are the relative exit velocity of extruded profile and the mandrel deflection, "the smaller-the-better type" should be applied. The *SN* ratio of this type can be calculated as:

$$SN = -10 \log \left[\frac{1}{n} \sum_{i=1}^{n} y_i^2 \right].$$

(1)

where n is the number of all the data points and y_i is the ith value of the relative exit velocity of extruded profile and the mandrel deflection.

3. Thermo-Mechanical Modelling

3.1. Basic Theories of Numerical Algorithm and Material Model

The arbitrary Lagrangian Eulerian (ALE) algorithm has been adopted for the hot extrusion simulation of complex profiles [9,12]. This is because the ALE method is a combination of Lagrangian and the Eulerian approach, in which the mesh distortion and the difficulty of the free surface tracking can be effectively avoided. In the other word, the mesh in such a formulation can move arbitrarily, i.e., the mesh is not attached to the material particle like the Lagrangian method and not fixed in the space like the Eulerian approach as well. Thus, the computational mesh of the ALE algorithm can even move in one direction and be fixed in the other direction. The mass conservation law, momentum and energy equations can be expressed as follows:

For mass conservation law,

$$\left. \frac{\partial \rho}{\partial t} \right|_{\zeta} + \frac{\partial \rho w_i}{\partial \xi_i} = 0.$$

(2)

For momentum equation,

$$\rho \left. \frac{\partial v_i}{\partial t} \right|_{\zeta} + \rho w_j \frac{\partial v_i}{\partial \xi_j} = \frac{\partial T_{ji}}{\partial \xi_j} + \rho f_i.$$

(3)

For energy equation,

$$\rho \left. \frac{\partial e}{\partial t} \right|_{\zeta} + \rho w_i \frac{\partial e}{\partial \xi_i} = T_{ji} \frac{\partial v_i}{\partial \xi_j} - \frac{\partial q_i}{\partial \xi_i}.$$

(4)

where ρ and T_{ji} are the density and the Lagranigain stress tensor respectively, w_i and w_j are the particle velocities in the referential coordinates, v_i is the particle velocity in the spatial coordinates. f_i is the body force per unit mass, and e is the internal energy per unit mass. The conservation equations for the mass, momentum and energy are necessary to be discretized and assembled into one system of the equations, which are calculated by using the iterative formulation.

In addition, a proper material constitutive relation is also important to describe the high-temperature deformation behavior of aluminum alloys. In this work, the typical Sellars-Tegart model is adopted to describe the extrusion deformation and yield the steady-state effective deviatoric flow stress as a function of parameters such as strain, strain rate, temperature, and so on [5]. It can be expressed as

$$\sigma = \frac{1}{\beta} \sinh^{-1} \left(\frac{Z}{A} \right)^{1/n}.$$

(5)

where σ is the flow stress, n is a stress exponent, A is a constant, β is the temperature-independent material parameter, and Z is the Zener-Hollomon parameter defined by

$$Z = \dot{\bar{\varepsilon}} e^{Q/RT}.$$

(6)

where $\dot{\bar{\varepsilon}}$ is the effective strain rate, Q is the activation energy, R is the universal gas constant, and T is the absolute temperature (K). In this work, the aluminum alloy 6060 and the H13 steel are chosen as the deformable work-piece and the tool material, respectively. Aluminum alloy has been one of widely used materials because of its relatively light weight and satisfactory strength properties [23]. In the past decade, more than 80% of the manufactured extruded aluminum products are based on the AA6xxx series alloys due to their high extrudability [24]. In particularly, aluminum alloy AA6060 is the most widely used for the productions of the extruded thin-walled tubes in real applications. The strength of the alloy is dependent on many aspects such as the amount of Mg and Si in the solid solution, and the size and distribution of the Mg_2Si precipitate particles. For a specific temperature, the flow stress of the studied AA6060 increases with the increase of strain rate. And for a fixed strain rate, the flow stress decreases apparently with the increase of temperature. The die material H13 steel with good abrasion resistance, hot hardness, and low sensitivity to heat checking can be subjected to drastic heating and cooling at a high process temperature. According to the above mentioned constitutive model, the corresponding material constitutive parameters of the studied AA6060 and the main physical properties of the H13 steel are listed in Table 2.

Table 2. Material parameters of AA6060 and H13.

Material	Q, J/mol	R, J/(mol·K)	A, s^{-1}	n	β, m^2/N
AA6060	1.44×10^5	8.314	5.91×10^9	3.515	3.46×10^{-8}
Material	E, GPa	λ, N/(s·C)	ρ, kg/m^3	υ	δ, N/(mm^2·C)
H13	210	24.3	7870	0.3	460

3.2. Process Modelling

The finite element simulation of the multi-output porthole extrusion process is performed with the commercial code HyperXtrude (version 10.0) [25]. It is one of the products particularly for the extrusion manufactures from Altair Engineering Company (Troy, MI, USA). HyperXtrude is a virtual press where the user can visualize the material flow and the temperature inside a particular die during the extrusion processes. It makes the necessary changes to ensure the balanced flow and eliminate the product defects. The HyperXtrude can help to reduce the cost of the die-trials. The mesh element size has effects on the simulation accuracy and the computing time cost. Generally, the smaller mesh size implies a large amount of the elements and the higher prediction accuracy but the longer computing time. Due to the studied complex thin-walled profile with only 0.7 mm thickness, the mesh elements in the bearing area should be refined. Here, the triangular prism element is used in the part of the die bearing and the profile (see Figure 7), while the tetrahedral element is adopted in the other parts. In order to balance the computing time cost and the element number and calculation accuracy, a quarter of the geometric model is used for the simulations due to the symmetry. The varying mesh density with the whole model is carried out in accordance with the extent of local deformation. During the multi-output porthole extrusion, the areas close to the die bearing will undergo severe shear deformations because the final shape of the profile is formed in the die bearing. Thus, five layers of the elements in the regions of the bearing and the profile are assigned during the meshing stage. Meanwhile, the relatively coarse meshes are performed in the other regions. It should be noted that the each simulation work takes about 20–22 h to run for the studied extrusion process.

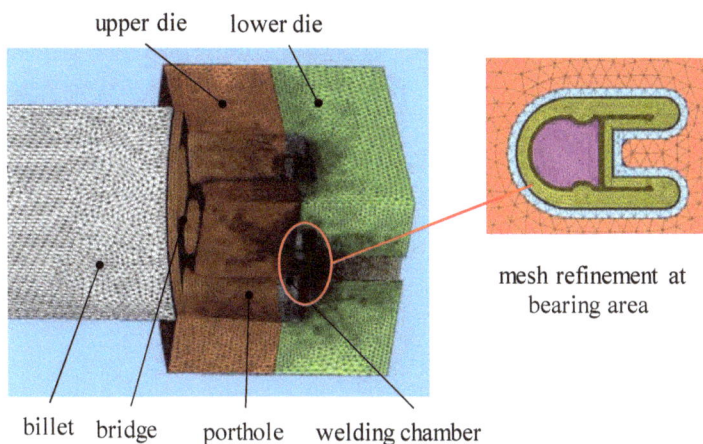

Figure 7. Finite element modelling of the multi-output porthole extrusion.

The initial billet used is 180 mm in diameter and 400 mm in length, and the extrusion ratio is up to 96.7. The initial temperature of the billet and the tools including dies and container are 480 °C and 430 °C, respectively. The extrusion ram speed is 5 mm/s. The heat convection and coefficient between the die and the billet material is 3000 W/m^2·C as well as the work converted to heat of 90%. In order to simplify the complex interfacial conditions ranging from full sticking at the die entrance to slipping at the end of the bearing, an empirical friction factor of 0.3 is prescribed in the whole model. Because the profile extrusion processes usually involve a high contact pressure, it is generally more appropriate to use the law of constant plastic shear friction. A friction factor of 0.3 was prescribed to simplify the complex interfacial conditions. The extrusion process parameters used in the simulations are given in Table 3.

Table 3. The process parameters of the multi-output porthole extrusion.

No.	Process Parameters	Values
1	Billet diameter, mm	180
2	Billet length, mm	400
3	Extrusion ratio	96.7
4	Ram speed, mm/s	5.0
5	Billet preheat, °C	480
6	Container initial temperature, °C	430
7	Heat convection coefficient, W/m^2·C	3000
8	Work converted to heat, %	90
9	Quantity of output	6

4. Results and Discussions

4.1. The Sources of Unbalanced Deformation

In order to explore the sources of unbalanced deformations of the aluminum alloy profile, two control strategies are proposed to obtain the optimal aluminum alloy extruded profile. Especially for the non-symmetrical profiles or even the multi-output porthole extrusion, it is more difficult to obtain the ideal shape profile after the hot extrusion processes. One of the sources of unbalanced deformation is the non-uniform relative exit velocity during the multi-output porthole extrusion process, as shown in Figure 8. It can be observed that the bearing length in the red areas should be increased to slow down the exit velocity. Thus, the bearing length in the blue area can be reduced to boost the exit material flow. In order to assure the dimension precision of the products and avoid various defects, such as twisting, bending, waving, cracks, etc., the outflow lengths (also known as bearing length or die land length) of the extrusion tools from each die orifice should be nearly identical. In the other word, all points in the cross section of the extruded profile should flow out of the die orifices with the same speed. Without such a uniform flow control, even simple profiles may not be extruded successfully. Of particular importance is the metal flow velocity at the die exit to judge whether this die will perform adequately in practice.

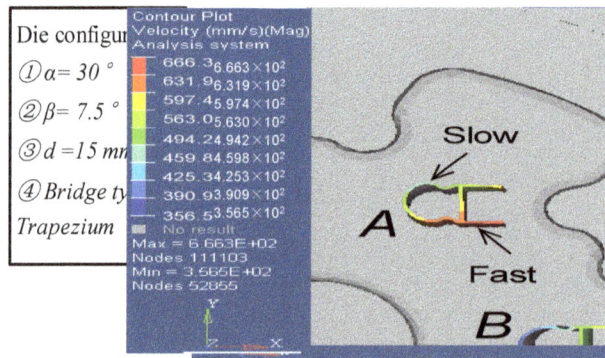

Figure 8. The distribution of the relative exit velocity.

To describe the metal flow velocity distribution at the die exit, the standard deviation of the velocity field in the axial direction (SDV) is introduced as

$$\text{SDV} = \sqrt{\frac{\sum_{i=1}^{m}(v_i - \bar{v})^2}{m}}. \tag{7}$$

The smaller value of the SDV, the better metal flow balance of the relative exit velocity or the extrusion quality. Therefore, the optimization of the bearing length and some key die features should be an efficient method to balance the exit material flow at the output side during the multi-output porthole extrusion process.

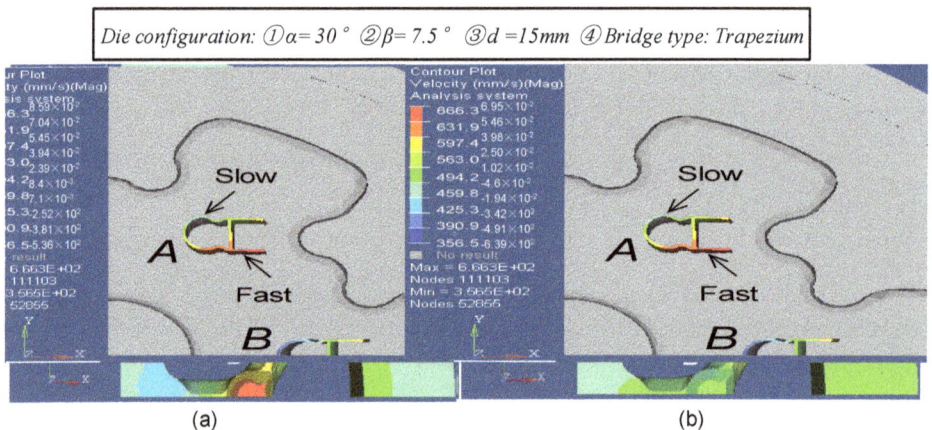

Figure 9. The mandrel deflections of the upper die: (**a**) *X*-displacement, (**b**) *Y*-displacement.

The other source is the mandrel deflection or the elastic deformation of the upper die during the multi-output extrusion process. Figure 9 shows the mandrel deflections of the upper die in one case of die configurations. The corresponding predicted maximum mandrel deflections (MMD) are listed in Table 4. The maximum wall thickness change of the extruded profile is up to 14% (0.0989/0.7). It is clear that the mandrel deflection should be considered at the design stage since it can cause a significant and undesirable wall thickness distribution of the final profile.

Table 4. The predicted mandrel deflections in one case of die configurations.

Displacement Magnitude	MMD-A (mm)	MMD-B (mm)
X-displacement (max.)	0.0703	0.0859
Y-displacement (max.)	0.0695	0
XY-displacement (max.)	0.0989	0.0859

In practice, the efficient control of the mandrel deflection is still treated by the know-how experience and the "try and error" experiments. Meanwhile, the predicted results by analytical method deviate far from the experimental ones since the complicated multi-factor forming conditions are difficult to be considered in the formulas. Besides the experimental and analytical methods, the optimization design of the die structures including the bearing length, the pocket and the welding chamber integrated with accurate prediction by means of thermo-mechanical modeling can be an alternative control strategy for the multi-output porthole extrusion process.

4.2. Analysis of Variance

Three levels of each parameter were determined and arranged according to the L9 orthogonal table of the Taguchi method as shown in Table 5. Generally, 81 cases should be examined if three levels of four parameters were fully arranged. However, only nine cases were examined using the Taguchi method. The SN ratios of the four shape parameters are presented in Figure 10. The optimum values of the β were not the same in the two cases and the SN ratio varied significantly. This means that the β has the highest effect on the mandrel deflection and the relative exit velocity of the extruded profile.

Table 5. The parameter conditions based on the L9 orthogonal array of the Taguchi method.

No.	α (°)	β (°)	*d* (mm)	Bridge Type	MMD-A (mm)	SDV-A (mm/s)
1	5	5	10	Brale	0.0989	10.256
2	15	7.5	15	Trapezium	0.0912	9.876
3	30	10	20	Round	0.0993	9.514
4	5	5	10	Round	0.0970	10.058
5	15	7.5	15	Brale	0.0905	9.879
6	30	10	20	Trapezium	0.1042	9.543
7	5	5	10	Trapezium	0.0992	10.119
8	15	7.5	15	Round	0.0918	9.894
9	30	10	20	Brale	0.1003	9.486

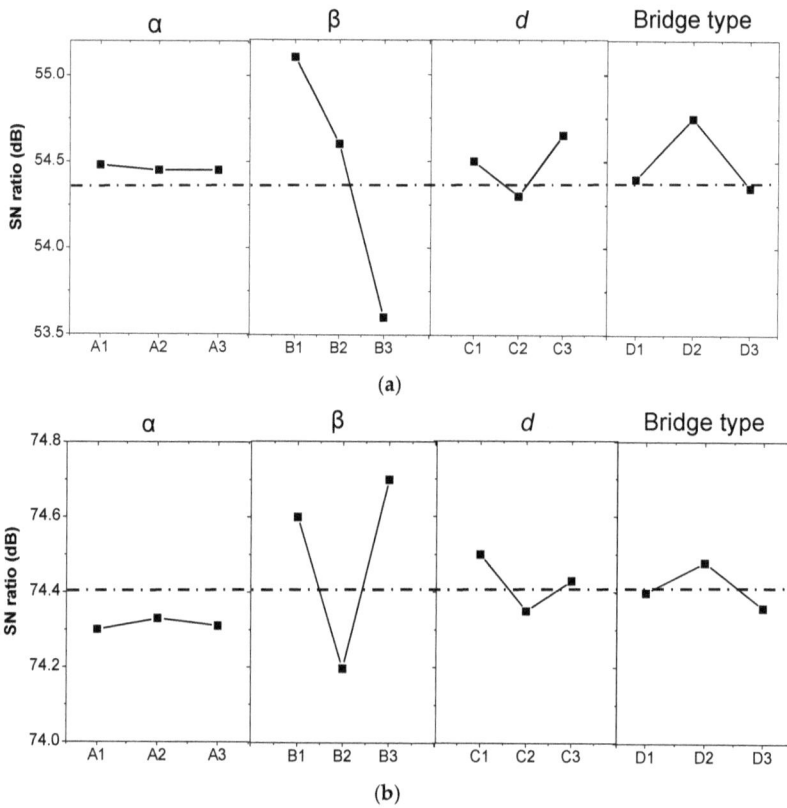

Figure 10. The SN (Signal-to-Noise ratio) ratio values of four shape parameters for (**a**) MMD (maximum mandrel deflections) and (**b**) SDV (axial direction).

Though the SN ratios indicated different values of α for the optimum value, the variation of α was very small. This means that the billet buckle angle has a slight effect on the mandrel deflection and/or the distortion of the extruded profile. It may be because the billet buckle angle α is far from the final extrusion process. Thus, we can choose any selected value of α for the optimum value. However, the billet buckle angle α in the upper die should play a considerable effect on the metal flow in the beginning of the multi-output extrusion process. For example, the flash overflow of material is prone to occur without the billet buckle angle α, as shown in Figure 11. The optimum value of the porthole bevel angle β has a significant effect on the mandrel deflection and the relative exit velocity in the multi-output porthole extrusion. The increase of the porthole bevel angle β decreases the mandrel deflection. However, the relative exit velocity decreases first and then increases with the increase of the porthole bevel angle β. The author attempts to explain that too large porthole bevel angle might cause the highlight of the direction change of the metal flow. Therefore, an appropriate porthole bevel angle is vital for the balance of the metal flow and the final profile distortion. The depth of the first welding chamber d seems to have effects but not very be obvious on the metal flow during the multi-output porthole extrusion. As for the effect of the bridge type, the trapezium of the bridge type seems to have the best chosen compared to the other types, because it can meet the demand the tool strength and the metal flow. The round type has the lowest of MMD because of the lower inlet bevel angle compared to the trapezium type. As for the brale type, the authors think the bridge strength is not strong enough for a high pressure extrusion.

Figure 11. The flash overflow of the material without billet buckle angle in the upper die.

Based on the above results, 30° of the billet buckle angle, 7.5° of the porthole bevel angle, and 15 mm of the depth of first welding chamber and the trapezium shape of the bridge type were determined as the optimum values for the mandrel deflection and the relative exit velocity, respectively. The maximum of the mandrel deflection at the location A and the relative exit velocity of the extruded profile having the optimum die feature parameters were predicted and validated experimentally.

4.3. Experimental Validation

The experiment equipment for the multi-output porthole extrusion process is a horizontal extrusion machine (maximum capacity 2000 Ton), namely SCHLOTMANN (NATIONAL MACHINERY EXCHANGE, INC., Newwark, NJ, USA). The main extrusion process parameters are listed in Table 3. The experimental observations of the upper die, the final extruded profiles and the distortion of the extruded profile are shown in Figure 12. It can be seen that the six-output extruded profiles have only a slight distortion at the head of extrusion and good surface quality. It is a successful industry application case for the presented multi-output porthole extrusion.

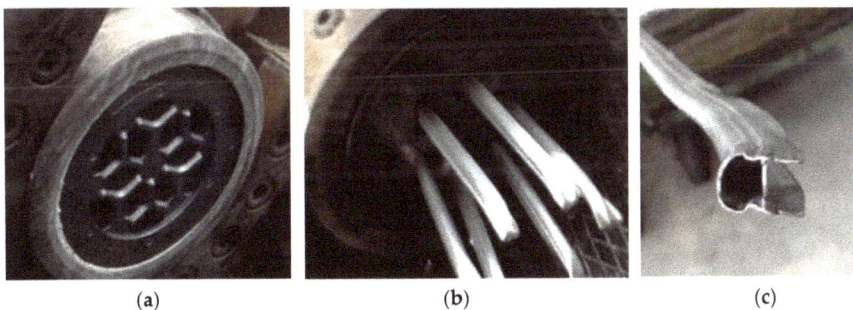

(a)   **(b)**   **(c)**

Figure 12. Experimental observations: (**a**) The upper die, (**b**) the extruded profiles, and (**c**) the nose end of the extruded profile with a slight shape distortion.

Figure 13 shows the comparison results of the wall thickness deviation along the section loci of the profile at the location A and the location B as previously shown in Figure 8, respectively. It can be seen that the predicted results at the location A are a little greater compared to the experimental ones. The maximum error occurred at the point 7 is about 0.012 mm. This may be due to the point 7 is the most far area to the center of the extrusion dies and has the largest moment of deformations.

The other relative high errors occur at the point 8 for both of the location A and B, which are the area of middle wall or cantilever. This might be the serious contraction at the middle cantilever after the cooling of the extrusion process. This phenomenon has not been predicted by the adopted numerical model. In addition, the wall thickness deviation of the extruded profile at the location B is smaller than that at the location A. This may be due to that the profile at the horizontal location B has a less mandrel deflection compared to the location A. It also can be seen that the wall thickness distribution of the profile at the location B is almost symmetric to the horizontal center line.

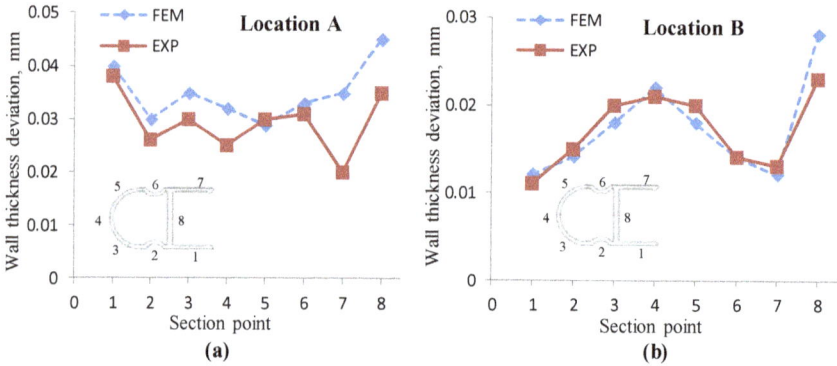

Figure 13. Comparison of wall thickness deviation between experimental and simulation results: (**a**) Location A and (**b**) Location B.

The maximum extrusion pressure recorded at the experimental extrusion machine is 205 Ton larger than that predicted 191.7 Ton. The force deviation between the predicted result and the experimental one is about 6.93%. This is a relative high prediction accuracy and acceptable for an industrial extrusion case with a complex section.

5. Conclusions

In this paper, a typical industrial extruded aluminum alloy thin-walled profile with a snap-fit channel produced by the proposed multi-output porthole extrusion process was addressed. The design scheme of the multi-output porthole extrusion die was introduced as well as the development of its thermo-mechanical model. In order to explore the sources of the undesirable deformations and the unbalanced metal flow, the non-uniform relative exit velocity of the extruded profile and the mandrel deflection were analyzed based on the simulation model. Meanwhile, the optimization design of four key die structures was performed by using the Taguchi method. The main conclusions of this study are as follows.

(1) The main structure design of the multi-output porthole extrusion die includes some novel features such as a circular pattern of the portholes with a dart-shape inlet bridge, a buckle angle in the inlet side of the upper die, a two-step welding chamber, and a non-uniform bearing length distribution. The presented die scheme is helpful to get a balance metal flow and reduce the forming load.

(2) Four die structure parameters: The billet buckle angle, the porthole bevel angle, the depth of welding chamber, and the bridge type were optimized by the Taguchi method and optical simulation. Their optimum values are 30° of the billet buckle angle, 7.5° of the porthole bevel angle, 15 mm of the depth of first welding chamber, and the trapezium shape of bridge types, respectively.

(3) The billet buckle angle α in the upper die should have a considerable effect on the metal flow in the beginning of the extrusion process rather than the metal flow control in the porthole

stage. The porthole bevel angle β has significant effects on the mandrel deflection and the relative exit velocity in the multi-output porthole extrusion. The increase of the porthole bevel angle β decreases the mandrel deflection. The depth of first welding chamber *d* and the bridge type seem to have effects that may not be obvious on the metal flow during the multi-output porthole extrusion

(4) In order to explore the sources of unbalanced deformations of aluminum alloy profile, two control strategies based on the simulation model, i.e., the optimal bearing length and the mandrel deflection compensation, are proposed to obtain a better metal flow and thin-walled aluminum alloy extruded profile.

(5) The results of experimental validation show that the maximum wall thickness deviation has only 0.012 mm for the profile with a 0.7 mm of main wall thickness, and the force deviation between the predicted result and the experimental one is about 6.93%. This is a relative high prediction accuracy and acceptable for an industrial extrusion case with a complex section.

Author Contributions: X.X. and J.L. conceived and designed the multi-output porthole extrusion die; X.X., G.V., and A.B.P. performed the experiments; X.X., and J.P. analyzed the data; X.X., G.V., J.L., and A.B.P. contributed reagents/materials/analysis tools; X.X. and J.L. wrote the main paper.

Funding: The project supports from the Portuguese Foundation of Science and Technology (SFRH/BPD/114823/2016, UID/EMS/00481/2013, CENTRO-01-0145-FEDER-022083), National Natural Science Foundation of China (No. 51705080), the Natural Science Foundation of Fujian Province (No. 2018J01761; No. 2018J01764), Open test funding project for valuable instruments in Fuzhou University (2018T015) and the Fuzhou Science and Technology Project (2016-G-67).

Acknowledgments: The authors wish to thank Portugal industrial companies Bi-silque S.A. and Extrusal S.A. for the assistance of all the extrusion experiments and funding support.

Conflicts of Interest: The authors declare no conflict of interest.

References

1. Berndt, N.; Frint, P.; Wagner, M.F.-X. Influence of Extrusion Temperature on the Aging Behavior and Mechanical Properties of an AA6060 Aluminum Alloy. *Metals* **2018**, *8*, 51. [CrossRef]
2. Pan, J.Y.; Xue, X. Numerical investigation of an arc inlet structure extrusion die for large hollow sections. *Int. Mater. Form.* **2018**, *11*, 405–416. [CrossRef]
3. Fan, X.; Chen, L.; Chen, G.; Zhao, G.; Zhang, C. Joining of 1060/6063 aluminum alloys based on porthole die extrusion process. *J. Mater. Process. Technol.* **2017**, *250*, 65–72. [CrossRef]
4. Gagliardi, F.; Ambrogio, G.; Filice, L. On the die design in AA6082 porthole extrusion. *CIRP Ann. Manuf. Technol.* **2012**, *61*, 231–234. [CrossRef]
5. Lof, J.; Blokhuis, Y. FEM simulations of the extrusion of complex thin-walled aluminum sections. *J. Mater. Process. Technol.* **2002**, *122*, 344–354. [CrossRef]
6. Fang, G.; Zhou, J.; Duszczyk, J. FEM simulation of aluminum extrusion through two-hole multi-step pocket dies. *J. Mater. Process. Technol.* **2009**, *209*, 1891–1900. [CrossRef]
7. Sinha, M.K.; Deb, S.; Das, R.; Dixit, U.S. Theoretical and experimental investigations on multi-hole extrusion process. *Mater. Des.* **2009**, *30*, 2386–2392. [CrossRef]
8. Sinha, M.K.; Deb, S.; Dixit, U.S. Design of a multi-hole extrusion process. *Mater. Des.* **2009**, *30*, 330–334. [CrossRef]
9. Zhang, C.; Zhao, G.; Chen, H.; Guan, Y.; Kou, F. Numerical simulation and metal flow analysis of hot extrusion process for a complex hollow aluminum profile. *Int. J. Adv. Manuf. Technol.* **2012**, *60*, 101–110. [CrossRef]
10. Sun, Z.C.; Cao, J.; Wu, H.L.; Yin, Z.K. Inhomogeneous deformation law in forming of multi-cavity parts under complex loading path. *J. Mater. Process. Technol.* **2018**, *254*, 179–192. [CrossRef]
11. Lee, J.M.; Kim, B.M.; Kang, C.G. Effects of chamber shapes of porthole die on elastic deformation and extrusion process in condenser tube extrusion. *Mater. Des.* **2005**, *26*, 327–336. [CrossRef]

12. Zhang, C.; Zhao, G.; Chen, H.; Guan, Y.; Cai, H.; Gao, B. Investigation on Effects of Die Orifice Layout on Three-Hole Porthole Extrusion of Aluminum Alloy 6063 Tubes. *J. Mater. Eng. Perform.* **2013**, *22*, 1223–1232. [CrossRef]

13. Gagliardi, F.; Ciancio, C.; Ambrogio, G. Optimization of porthole die extrusion by Grey-Taguchi relational analysis. *Int. J. Adv. Manuf. Technol.* **2018**, *94*, 719–728. [CrossRef]

14. Bakker, A.J.D.; Werkhoven, R.J.; Sillekens, W.H.; Katgerman, L. The origin of weld seam defects related to metal flow in the hotextrusion of aluminum alloys EN AW-6060 and EN AW-6082. *J. Mater. Process. Technol.* **2014**, *214*, 2349–2358. [CrossRef]

15. Bakker, A.J.D.; Katgerman, L.; Zwaag, V.D.S. Analysis of the structure and resulting mechanical properties of aluminum extrusions containing a charge weld interface. *J. Mater. Process. Technol.* **2016**, *229*, 9–21. [CrossRef]

16. Zhang, C.; Dong, Y.; Wang, C.; Zhao, G.; Chen, L.; Sun, W. Evolution of transverse weld during porthole extrusion of AA7N01 hollow profile. *J. Mater. Process. Technol.* **2017**, *248*, 103–114. [CrossRef]

17. Yu, J.; Zhao, G.; Cui, W.; Zhang, C.; Chen, L. Microstructural evolution and mechanical properties of welding seams in aluminum alloy profiles extruded by a porthole die under different billet heating temperatures and extrusion speeds. *J. Mater. Process. Technol.* **2017**, *247*, 214–222. [CrossRef]

18. Gamberoni, A.; Donati, L.; Reggiani, B.; Haase, M.; Tomesani, L.; Tekkaya, A.E. Industrial Benchmark 2015: Process monitoring and analysis of hollow EN AW-6063 extruded profile. *Mater. Today Proc.* **2015**, *2*, 4714–4725. [CrossRef]

19. Selvaggio, A.; Kloppenborg, T.; Schwane, M.; Hölker, R.; Jäger, A.; Donati, L.; Tomesani, L.; Tekkaya, A.E. Extrusion benchmark 2013—Experimental analysis of mandrel deflection, local temperature and pressure in extrusion dies. *Key Eng. Mater.* **2013**, *585*, 13–22. [CrossRef]

20. Selvaggio, A.; Donati, L.; Reggiani, B.; Haase, M.; Dahnke, C.; Schwane, M.; Tomesani, L.; Tekkaya, A.E. Scientific Benchmark 2015: Effect of choking and bearing length on metal flow balancing in extrusion dies. *Mater. Today Proc.* **2015**, *2*, 4704–4713. [CrossRef]

21. Chen, P.C.; Chen, Y.C.; Pan, C.W.; Li, K.M. Parameter Optimization of Micromilling Brass Mold Inserts for Microchannels with Taguchi Method. *Int. J. Precis. Eng. Manuf.* **2015**, *16*, 647–651. [CrossRef]

22. Maheedhara, R.G.; Diwakar, R.V.; Satheesh, K.B. Experimental Investigation on radial ball bearing parameters ssing Taguchi Method. *J. Appl. Comput. Mech.* **2018**, *6*, 69–74.

23. Esat, V.; Darendeliler, H.; Gokler, M.I. Finite element analysis of springback in bending of aluminum sheets. *Mater. Des.* **2002**, *23*, 223–229. [CrossRef]

24. Cai, M.; Field, D.P.; Lorimer, G.W. A systematic comparison of static and dunamic aging of two Al-Mg-Si alloys. *Mater. Sci. Eng. A* **2004**, *373*, 64–71. [CrossRef]

25. Altair Hyperworks, Material Database. Altair Engineering, Inc., 2018. Available online: https://altairhyperworks.com/ (accessed on 15 June 2018).

metals

MDPI

Article

Effect of Process Parameters on Fatigue and Fracture Behavior of Al-Cu-Mg Alloy after Creep Aging

Lihua Zhan [1,2,*], **Xintong Wu** [1,2], **Xun Wang** [1,2], **Youliang Yang** [1,2], **Guiming Liu** [1,2]
and Yongqian Xu [1,3]

[1]   State Key Laboratory of High-Performance Complex Manufacturing, Central South University,
     Changsha 410083, China; wuxintong@csu.edu.cn (X.W.); doudouaitutu1112@163.com (X.W.);
     133711090@csu.edu.cn (Y.Y.); 163712126@csu.edu.cn (G.L.); yongqian.xu@csu.edu.cn (Y.X.)
[2]   School of Mechanical and Electrical Engineering, Central South University, Changsha 410083, China
[3]   Light Alloy Research Institute, Central South University, Changsha 410083, China
*   Correspondence: yjs-cast@csu.edu.cn

Received: 2 April 2018; Accepted: 21 April 2018; Published: 26 April 2018

Abstract: A set of creep aging tests at different aging temperatures and stress levels were carried out for Al-Cu-Mg alloy, and the effects of creep aging on strength and fatigue fracture behavior were studied through tensile tests and fatigue crack propagation tests. The microstructures were further analyzed by using scanning electron microscopy (SEM) and transmission electron microscopy (TEM). The results show that temperature and stress can obviously affect the creep behavior, mechanical properties, and fatigue life of Al-Cu-Mg alloy. As the aging temperature increases, the fatigue life of alloy first increases, and then decreases. The microstructure also displays a transition from the Guinier-Preston-Bagaryatsky (GPB) zones to the precipitation of S phase in the grain interior. However, the precipitation phases grow up and become coarse at excessive temperatures. Increasing stress can narrow the precipitation-free zone (PFZ) at the grain boundary and improve the fatigue life, but overhigh stress can produce the opposite result. In summary, the fatigue life of Al-Cu-Mg alloy can be improved by fine-dispersive precipitation phases and a narrow PFZ in a suitable creep aging process.

Keywords: Al-Cu-Mg alloy; creep aging; fatigue fracture behavior; microstructure

---

1. Introduction

Creep aging forming is a kind of forming method using the combination of the creep deformation of metal and the aging of aluminum alloy, which is mainly used for manufacturing aircraft wing panels and other integral panel components. This method is of better forming precision and repeatability than that of conventional plastic forming, which reduces the possibility of material fracture in the process and the residual stress in the components [1–5]. In recent years, the research on the formation of creep aging has mainly focused on how to improve microstructures, mechanical properties, or forming precision. Ho et al. [6] presented a creep damage constitutive equation for creep aging through forming a springback simulation that considered precipitation phases change in view of the 7010 aluminum alloy. Zhan [7] improved on that model, and built a set of constitutive models that can simulate the change of creep strain, precipitation phase, dislocation strengthening, solid solution strengthening, aging reinforcement, and material properties of the forming process in combination with the creep unidirectional tensile test. Xu et al. [8] compared the precipitation behaviors of 2124 aluminum alloy under several aging forming conditions, finding that the key mechanism of the generation and control of the precipitation orientation effect of 2124 aluminum alloy lies in the effect of dislocation. Based on creep aging experimental data, Yang et al. [9] presented a constitutive modeling and springback simulation for 2524 aluminum alloy that was much closer to experimental results. These research studies have been gradually applied to manufacturing, which makes it possible to transform high performance aluminum alloy into complex components by creep aging forming. However, with the development of the aerospace industry, the components are faced with more complex and extreme service conditions, which require that the materials have not only good strength and toughness, but also excellent comprehensive performance.

The importance of these problems was recognized, and more studies on the service performance of aluminum alloy were carried out. Chen et al. [10] reported on the effects of inclusions, grain boundaries, and grain orientations on the fatigue crack initiation and propagation behaviors of 2524 aluminum alloy, finding that coarse inclusion particles drastically accelerate local crack growth rates and that fatigue crack shows a strong tendency to propagate transgranularly grains with high Schmid factors. Siddiqui [11], Liu [12], and Chen et al. [13] studied the effects of different aging treatments on the fatigue behavior of aluminum alloy in various environments. They focused on the relationship between microstructures and fatigue behavior, and proved that fatigue crack propagation was influenced by the aging time and temperature, which indicated that the fatigue with aging time resistance was linked to the vacancies that were assisted by the diffusion mechanism and dislocation movement. Although the effect of aging temperature on the comprehensive performance of aluminum alloy has been demonstrated in the last decades, little attention has been paid to the fatigue behavior of aluminum alloy after creep aging. At present, studies about creep aging mainly concentrate on finding out the creep aging strengthening behavior (mechanical property) of materials under different process parameters, regarding the peak strength of material as the basis of determining the best technological system of creep aging. Zhou [14] and Chen [15] studied the effects of aging and creep aging on the mechanical properties, fatigue crack growth performance, and exfoliation corrosion of aluminum alloy.

In this work, a widely used Al-Cu-Mg alloy for aerospace was used as the research object [16–19], and the microstructures were analyzed by scanning electron microscopy (SEM), and transmission electron microscopy (TEM). The purpose of this paper is to analyze the effects of different creep aging parameters on creep behavior, mechanical property, and fatigue fracture behavior of aluminum alloy. This work could provide basic technological parameters and theory for the manufacturing of large components so as to improve the development of creep aging.

2. Experimental

2.1. Samples and Heat Treatment

The experimental materials used were supplied in the form of homogenized rolled sheets, whose chemical composition is listed in Table 1. The initial state is T3, which refers to cold processing after solution treatment followed by natural aging and then a basically stable state. The samples were taken along the longitudinal (L) direction, which refers to GB/T2039-2012 (metallic materials—uniaxial creep testing method in tension), as shown in Figure 1. Figure 2 shows the stress–strain curve of the initial sample. The Young modulus, tensile strength, yield strength, and elongation of the initial sample are as follows: 69.2 ± 0.8 GPa, 460.3 ± 3.4 MPa, 307.4 ± 4.8 MPa, $21.6 \pm 0.6\%$ (Five horizontal samples were taken). Conventional artificial aging treatment and creep aging treatment were conducted on a plate, for which a RMT-D10 electron creep slackness tester (Zhuhai SUST Electrical Equipment Co., Ltd., Zhuhai, China) was adopted for creep aging. The temperature of creep aging was selected at 433 K, 453 K, and 473 K respectively. The creep stress was selected as 0 MPa, 180 MPa, and 210 MPa, respectively, and the aging time was 12 h.

Table 1. Main chemical composition of Al–Cu–Mg alloy (mass fraction, %).

Cu	Mg	Mn	Fe	Zn	Ti	Cr	Si	Al
4.26	1.36	0.57	0.037	0.024	0.01	0.002	0.089	Bal.

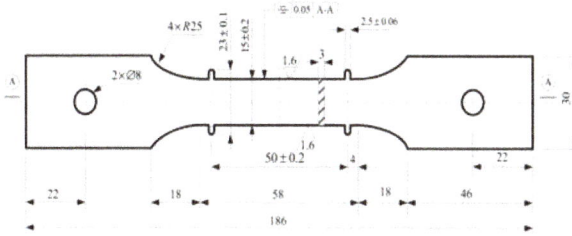

Figure 1. Standard creep aging sample (unit: mm).

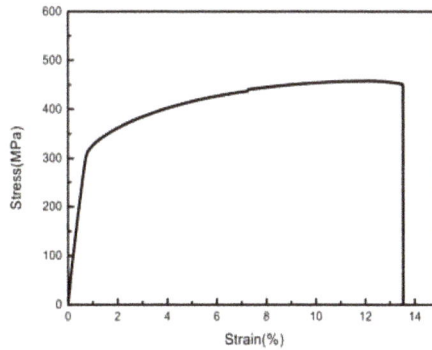

Figure 2. Stress-strain curve of the initial sample.

2.2. Tensile and Fatigue Testing

Tensile testing was carried out for the materials of different creep process parameters at room temperature on the E45 type testing machine (MTS Systems Corporation, Eden Prairie, MN, USA) with 2 mm/min loading speed. All of the fatigue tests were conducted through sine-wave loading at a stress ratio (R = $\sigma_{min}/\sigma_{max}$) of 0.1 (30/300 MPa) with a loading frequency of 10 Hz on an MTS810 tester machine (MTS Systems Corporation, Eden Prairie, MN, USA) fatigue tester at room temperature in a laboratory air environment. For each state, five horizontal samples were taken.

2.3. Microstructural Analysis

Fatigue fracture surfaces of the samples that were cyclically deformed in fatigue testing were analyzed by a TESCAN scanning electron microscope (Tescan company, Brno, Czech) so as to study the production and expansion of cracks. Samples suitable for a transmission electron microscope (TEM) were thin 0.8 mm slices that were electropolished by using twin-jet equipment with a voltage of 15 V in a 30% nitric and 70% methanol mixture acid at approximately $-30\,^{\circ}$C, after which the slices were cleaned in ethanol at room temperature for at least 5 min. These were then examined on a Tecnai G^2 20 TEM (United States FEI limited liability company, Hillsboro, OR, USA) machine operated at 200 kV.

3. Results and Discussion

3.1. Creep Behavior of Alloy

As shown in Figure 3, the curves of Al-Cu-Mg alloy were measured under different aging temperature and stress levels after 12 h of aging. The creep and steady-state creep rate at different aging temperature and stress levels are listed in detail in Table 2. From Figure 3 and Table 2, at the same temperature, the creep strain and steady-state creep rate rise with the increase of stress. At 180 $^{\circ}$C, the creep strain is 0.145% under a stress level of 180 MPa after 12 h of aging, while the creep increases to 0.169% under a stress level of 210 MPa. Moreover, the steady-state creep was also improved from 7.58×10^{-3} to 8.42×10^{-3} with the changed stress. At the same stress, the creep strain and steady-state creep rate also grow with the rise in temperature. When the stress level is 180 MPa, the creep strain is 0.068% after 12 h of aging at 160 °C. When the temperature increases to 200 °C, the creep strain is 0.235%, which is 3.5 times that at 160 °C. Meanwhile, the steady-state creep rate (1.033×10^{-2}) at 200 °C is 3.1 times that at 160 °C (3.3×10^{-3}).

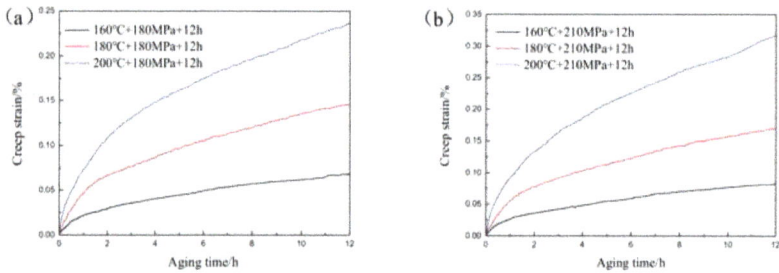

Figure 3. Creep curves of Al-Cu-Mg alloy under different aging temperature and stress levels: (a) 180 MPa; (b) 210 MPa.

Table 2. Creep strain and steady-state creep rate of Al-Cu-Mg alloy under different aging temperature and stress levels.

Temperature/°C	Stress/MPa	Creep Strain/%	Steady-State Creep/s^{-1}
160	180	0.068	3.3×10^{-3}
	210	0.082	4.02×10^{-3}
180	180	0.145	7.58×10^{-3}
	210	0.169	8.42×10^{-3}
200	180	0.235	1.033×10^{-2}
	210	0.315	1.46×10^{-2}

3.2. Mechanical Properties of Alloy

The effects of aging temperature and stress on the mechanical properties of alloy were studied by using an orthogonal test. The experimental method and results are summarized in Table 3. The results suggest that aging temperature has an obvious influence on the mechanical properties of the alloy. Tensile strength and yield strength first increases, and then decreases while the aging temperature is rising.

Table 3. Mechanical properties of the alloy after creep aging.

Temperature/°C	Stress/MPa	Tensile Strength/MPa	Yield Strength/MPa	Elongation/%
160	0	482.6 ± 2.5	331.6 ± 3.4	19.9 ± 0.3
160	180	497.5 ± 4.1	362.4 ± 5.6	21.6 ± 0.5
160	210	496.6 ± 3.0	358.6 ± 3.6	21.6 ± 0.4
180	0	495.6 ± 2.7	443.4 ± 3.4	10.1 ± 0.4
180	180	504.2 ± 4.5	463.2 ± 4.8	10.1 ± 0.3
180	210	501.4 ± 2.8	461.1 ± 3.1	10.4 ± 0.3
200	0	465.6 ± 3.6	372.6 ± 4.2	8.8 ± 0.5
200	180	472.7 ± 3.5	417.3 ± 4.6	8.9 ± 0.4
200	210	475.2 ± 2.1	419.9 ± 2.6	8.9 ± 0.3

3.3. Effect of Aging Temperature on Fatigue Fracture of Alloy

Figure 4 shows the fatigue life of Al-Cu-Mg alloy under a constant maximum stress of 180 MPa after 12 h of creep aging at different aging temperatures. It can be seen that the fatigue life of the alloy first increases, and then decreases with the increase of aging temperature under the same stress, and the highest is 90,971 cycles at 180 °C. Meanwhile, the fatigue life of the alloy at 200 °C is lower than that at 160 °C.

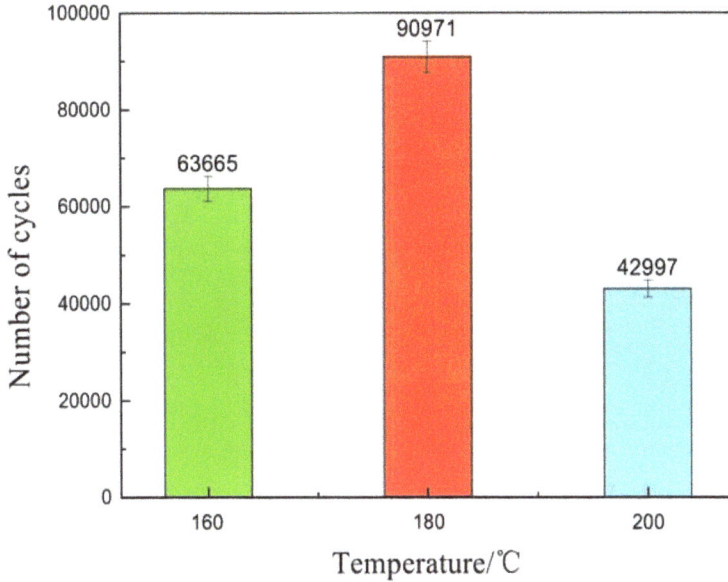

Figure 4. The fatigue life of the alloy under a constant stress of 180 MPa after 12 h of creep aging at different aging temperatures.

Figures 5–7 show the fatigue fracture appearance of Al-Cu-Mg alloy under a constant stress of 180 MPa after 12 h of aging at 160 °C, 180 °C, and 200 °C. These figures demonstrate that the fatigue surfaces can be divided into three regions: fracture nucleation, stable fatigue crack growth, and final fracture. In the fracture nucleation, the main cracks, some small secondary cracks, torn edges, and quasi-cleavage present a river pattern. The fracture has crystallographic characteristics, reflecting crack extensions along the different crystal planes and second-phase particles [20]. The fatigue life of metal materials mainly includes the fatigue cycles of fatigue crack and fatigue crack growth. Therefore, the larger the area of the fatigue source zone and stable fatigue crack growth, the greater the fatigue life. In addition, during the fatigue crack growth, a microtrace named fatigue striation appeared on the fatigue surface, which can be regarded as a fatigue cycle of the alloy. As a result, a narrower fatigue strip can also indicate that the fatigue life at 180 °C is higher.

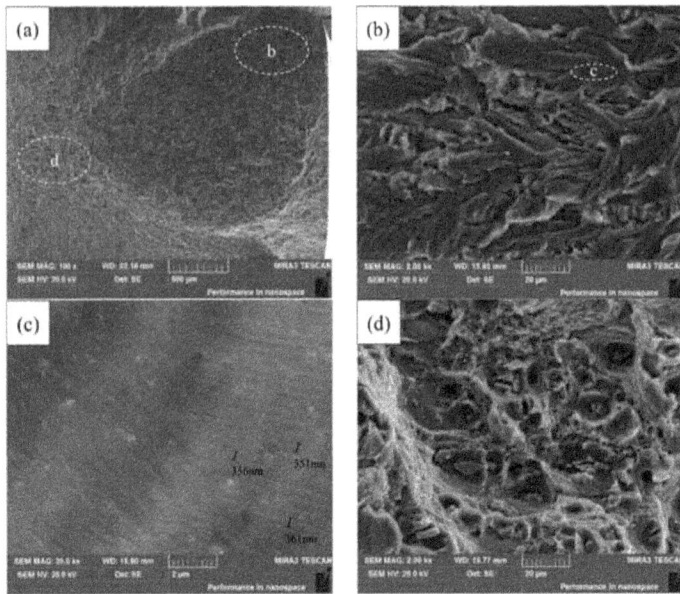

Figure 5. The fatigue fracture surfaces of Al-Cu-Mg alloy under 180 MPa after 12 h of creep aging at 160 °C: (**a**) full view; (**b**) fracture nucleation; (**c**) stable fatigue crack growth; (**d**) final fracture.

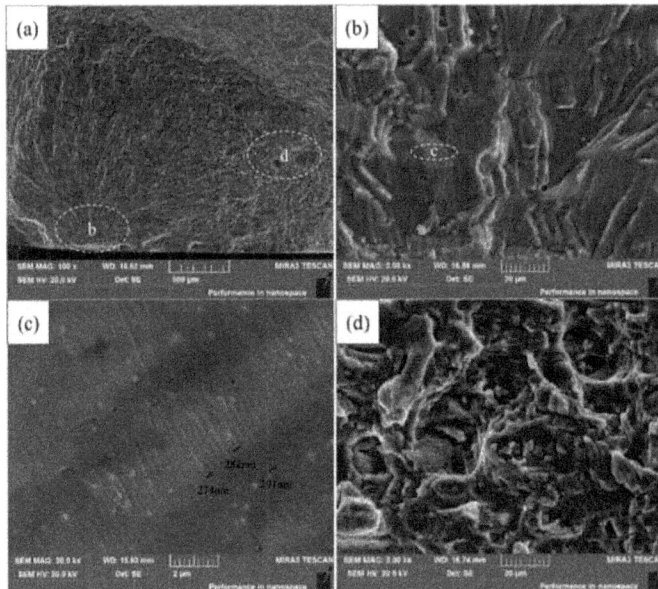

Figure 6. The fatigue fracture surfaces of Al-Cu-Mg alloy under 180 MPa after 12 h creep aging at 180 °C: (**a**) full view; (**b**) fracture nucleation; (**c**) stable fatigue crack growth; (**d**) final fracture [21].

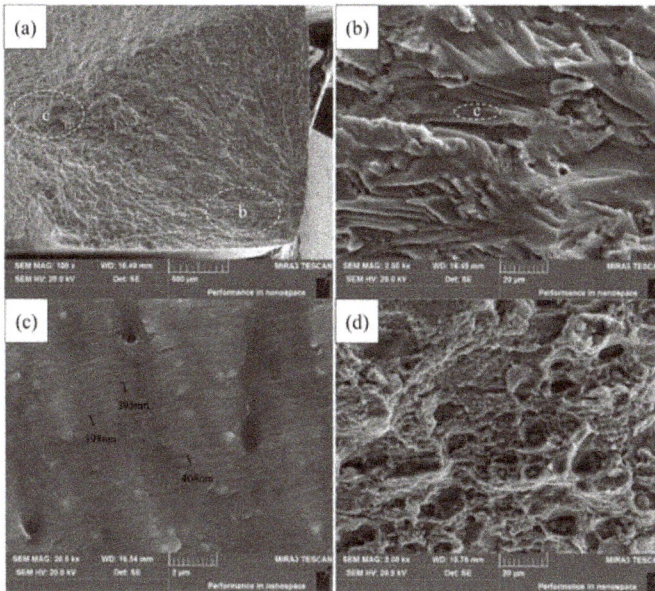

Figure 7. The fatigue fracture surfaces of Al-Cu-Mg alloy under 180 MPa after 12 h of creep aging at 200 °C: (**a**) full view; (**b**) fracture nucleation; (**c**) stable fatigue crack growth; (**d**) final fracture.

From Figure 5, it can be seen that the area of fatigue crack and fatigue crack growth is large under a constant stress of 180 MPa after 12 h of creep aging at 160 °C. Figure 6 shows the largest area of fatigue crack and fatigue crack growth at 180 °C, and its fatigue striation is more compact and regular than other conditions. While the creep aging temperature rises to 200 °C, in Figure 7, the fatigue crack and fatigue crack growth zone of the alloy is the smallest. Its fatigue striation is accompanied by some microcracks, and is not as clear and compact as other conditions. In conclusion, with the increase of the creep aging temperature, the area of fatigue crack and fatigue crack growth first increases, and then decreases under the same stress. The distance between the fatigue striation changes according to the same rule. These indicate that the fatigue life first increases and then decreases with the increase of temperature, which is consistent with the data obtained from previous fatigue life tests.

3.4. Effect of Stress on Fatigue Fracture of Alloy

Figure 8 shows the fatigue life of Al-Cu-Mg alloy under different stress after 12 h of creep aging at 180 °C. The fatigue life of alloy is only 37,510 cycles after 12 h of aging without stress, while it is obviously improved after stress is applied. However, the fatigue life decreases as the stress further increases.

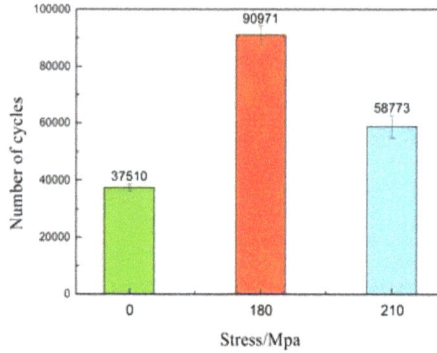

Figure 8. The fatigue life of the alloy under different stress levels after 12 h of creep aging at 180 °C.

Figures 9 and 10 show the fatigue fracture appearance of Al-Cu-Mg alloy under different stress levels (0 MPa, 210 MPa) after 12 h of aging at 180 °C. It can be seen in Figure 9 that the area of fatigue crack and fatigue crack growth is small, but the area of final fracture is large. Moreover, there are some slender microcracks at the fatigue striation on the surface, which is quite different from that observed previously. These cracks can significantly accelerate the fatigue process and reduce the fatigue life. When the stress rises to 210 MPa, the area of fatigue crack and fatigue crack growth is larger than that of 0 MPa, as presented in Figure 10. Furthermore, the fatigue striation is also narrower, and there is no microcrack. Compared with 0 MPa or 210 MPa, the fatigue striation of 180 MPa is the narrowest, as revealed in Figure 6.

In summary, stress can significantly improve the fatigue life of alloy during aging. However, with the stress further increasing, the fatigue life of alloy is reduced on the grounds of the data obtained from previous fatigue life tests.

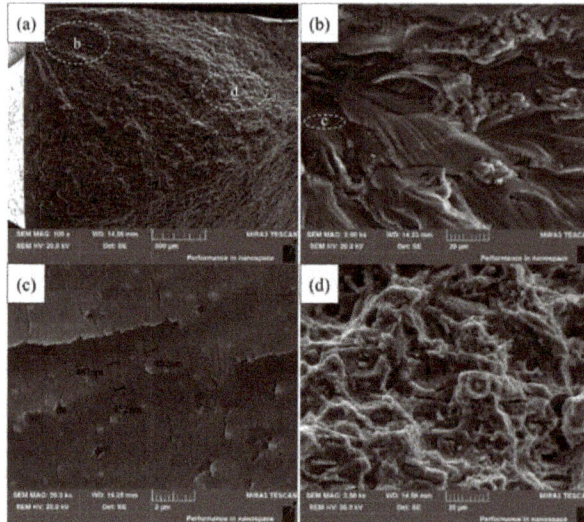

Figure 9. The fatigue fracture surfaces of Al-Cu-Mg alloy under 0 MPa after 12 h of creep aging at 180 °C: (**a**) full view; (**b**) fracture nucleation; (**c**) stable fatigue crack growth; (**d**) final fracture.

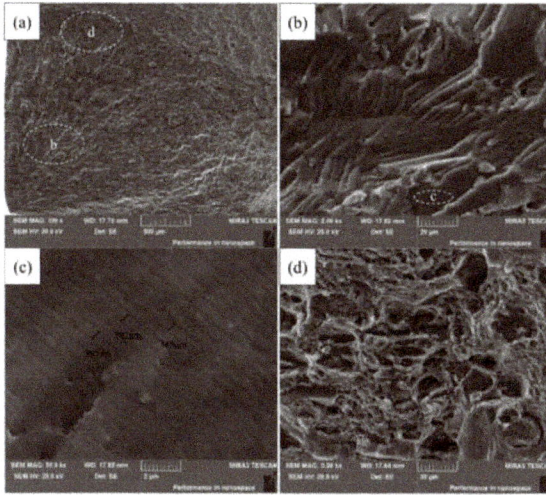

Figure 10. The fatigue fracture surfaces of A-Cu-Mg alloy under 210 MPa after 12 h of creep aging at 180 °C: (**a**) full view; (**b**) fracture nucleation; (**c**) stable fatigue crack growth; (**d**) final fracture.

3.5. Relationship between Microstructure and Fatigue Fracture Behavior of Alloy

3.5.1. Effect of Aging Temperature

Figure 11 shows the precipitation characteristics of Al-Cu-Mg alloy under 180 MPa stress after 12 h of creep aging at 160 °C, 180 °C, and 200 °C. At an aging temperature of 160 °C, there are Guinier-Preston-Bagaryatsky (GPB) zones in the grain interior under 180 MPa stress after 12 h. There are also rod-like precipitation phases in the grain interior, according to the relevant report [22]; this is called the T phase ($Al_{20}Cu_2Mn_3$), which is stable and forms during homogenization treatment or hot rolling. Meanwhile, dislocations appear in the grain interior due to long-term creep aging. There is neither a precipitation-free zone (PFZ) nor precipitation phase segregation at grain boundaries. Then, with the creep aging temperature rising to 180 °C, there are a large number of acicular S phases dispersed in the grain interior, strongly impeding the movement of dislocations, as presented in Figure 11c. At the same time, coarse precipitates are observed around the grain boundaries without PFZ. When the creep aging temperature rises to 200 °C, the fine phase S transforms to coarse phase S. Meanwhile, the phases grow up, and the grain boundary segregation occurs to form a continuous chain. The 105 nm wide PFZ forms at the grain boundaries, because the solute atoms preferentially precipitate at the grain boundaries.

During creep aging, more fine and dispersed precipitation phases are formed under the 180 °C condition than the others. Moreover, the large volume fraction of precipitation phases make the dislocation movement become more difficult, so it has a higher strength and greater fatigue crack propagation resistance [23]. Furthermore, with the temperature rising, precipitation phases also appear at the grain boundary, and finally, PFZ forms. In conclusion, the increasing of temperature will promote the growth of precipitation phases and PFZ.

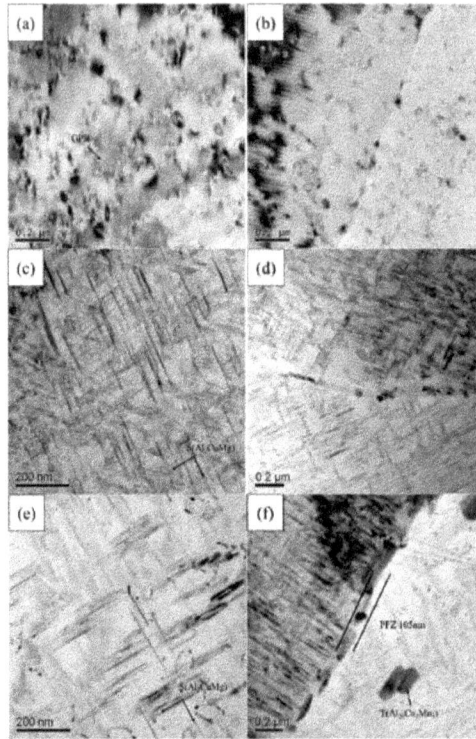

Figure 11. TEM micrographs of Al-Cu-Mg alloy under 180 MPa after 12 h of creep aging at different temperature: (**a,b**) 160 °C; (**c,d**) 180 °C; (**e,f**) 200 °C.

3.5.2. Effect of Stress

Figure 12 shows the TEM of Al-Cu-Mg alloy under different stress levels after 12 h of creep aging at 180 °C. There are small and dispersed acicular S phases in the grain interior, when stress is 0 MPa. The number of the precipitation phases is less compared with the 180 MPa condition because of the addition of stress. Stress deforms the alloy and increases dislocations in the grain interior, which provides a favorable nucleation position for the heterogeneous nucleation of solute atoms, thus accelerating the precipitation phases, which then precipitate more quickly. Meanwhile, there are precipitation phases nucleating at grain boundaries and a PFZ about 125 nm wide appearing under the stress of 0 MPa. However, compared with the stress at 180 MPa, there is neither a PFZ nor a precipitation phase. This is because the existence of stress brings about a large number of dislocations in the grain interior during the aging process, promoting precipitation phases precipitating at these dislocations, and making phases precipitate with uniform driving force. When stress is 210 MPa, the size of the precipitation phases in the grain interior is larger than that at 180 MPa, due to the increasing stress, which advances the precipitation process.

At the same temperature, the fatigue life of the alloy under 0 MPa is shorter than that after creep aging. This is because, on the one hand, stress can generate a part of plastic deformation, which increases dislocation density in the grain interior during the aging process, causing precipitation phases to precipitate more rapidly. On the other hand, the precipitating driving force in the grain interior and grain boundaries becomes more uniform due to the application of stress, so the PFZ narrows or disappears. These will reduce the stress concentration at grain boundaries, increasing the fatigue life of the alloy. However, overhigh stress brings about an improvement of the precipitation process and induces the appearance of a PFZ so that the fatigue life of the alloy declines.

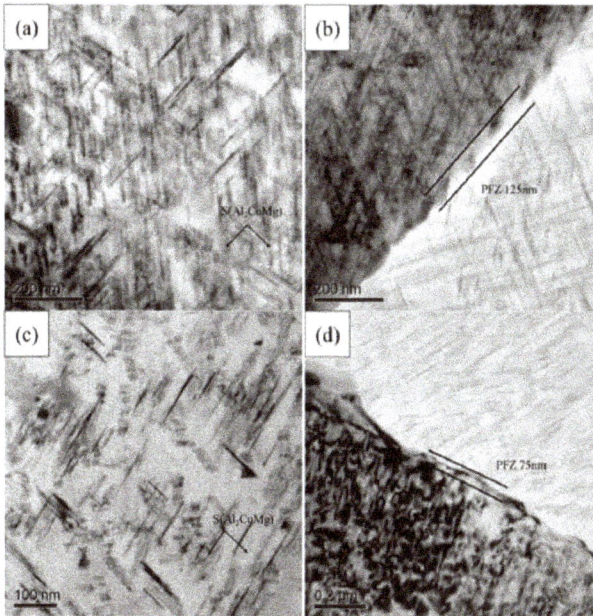

Figure 12. TEM micrographs of Al-Cu-Mg alloy under different stress levels after 12 h of creep aging at 180 °C: (**a**,**b**) 0 MPa; (**c**,**d**) 210 MPa [21].

There are many factors that affect the fatigue behavior of Al-Cu-Mg alloy, such as fatigue conditions, surface state, dimensions, and materials. The main purpose of this study is to research the effect of creep aging parameters on the fatigue fracture of Al-Cu-Mg alloy. The creep aging parameters (aging temperature, stress level) can obviously change the aging precipitation characteristics and microstructure before they ultimately affect the fatigue fracture behavior. The precipitation sequence of Al-Cu-Mg alloy used in this paper has generally been believed: supersaturated solid solution (SSSS) → solute clusters → Guinier–Preston–Bagaryatsky (GPB) → S [24–26]. The effects of aging precipitation characteristics on the fatigue properties of the alloy is reflected in three aspects [27,28]:

(1) During the creep aging process, the strengthening phases in the grain interior can precipitate under the combined impact of stress and temperature. These precipitation phases in the matrix impede the dislocations movement and improve the strength and anti-plastic deformation of alloy until they eventually reduce the deformation damage during the fatigue process and prolong fatigue life.

(2) The precipitation phases in the grain interior are semi-coherent or incoherent with the matrix, causing dislocations passing round the precipitation phases and leaving a dislocations loop. It makes dislocations unable to slip repeatedly in grains, which leads to the dislocations easily slipping to the grain boundary. The precipitation phase is not conducive to the crack growth and the reduction of deformation damage.

(3) The precipitation-free zone (PFZ) and coarse particles along the grain boundaries will reduce the strength of grain boundaries, so that the dislocations are more likely to slip at the grain boundary. These can increase the stress concentration of grain boundaries, forming cracks easily along the grain boundaries.

4. Conclusions

In this paper, the creep behavior, mechanical properties, microstructure, and fatigue crack behavior of an Al-Cu-Mg alloy were studied.

(1) With the increase of temperature and stress, the creep strain of the alloy is improved. As the temperature rises, the tensile strength and yield strength first increase, and then decrease. The mechanical properties of alloy are improved after creep aging, but stress has a weak effect on the mechanical properties.

(2) The process parameters can obviously impact the fatigue life of the alloy. The fatigue life first increases, and then decreases with the rise of both temperature and stress. Both overhigh temperature and stress can shorten the fatigue life. The best fatigue life is 90971 cycles at 180 °C under 180 MPa in this study.

(3) The precipitation characteristics of alloy are the main reason for the fatigue life change of the alloy. When the precipitation phases are fine and dispersed, the strength and anti-plastic deformation of the alloy are improved, and fatigue life increases. However, when the precipitation phases become coarse, the precipitation phases cannot be cut by dislocation, so the fatigue life dropped. Besides, the PFZ at the grain boundary is a weak part of alloy, where the fatigue life decreases with a wide PFZ.

Author Contributions: Lihua Zhan and Xintong Wu proposed the original project and supervised the investigation. Xintong Wu, Youliang Yang and Guiming Liu performed testing and characterization. Xintong Wu and Xun Wang performed data analysis and wrote the paper. Yongqian Xu contributed consultation, data analysis.

Acknowledgments: The authors gratefully acknowledge Financial Support of the National Science Foundation of China (No. 51675538) and the National Basic Research Program of China (No. 2014CB046502), for funding the work reported.

Conflicts of Interest: The authors declare no conflict of interest.

References

1. Zhan, L.; Lin, J.; Dean, T.A. A review of the development of creep age forming: Experimentation, modelling and applications. *Int. J. Mach. Tool Manuf.* **2011**, *51*, 1–17. [CrossRef]

2. Xu, Y.; Zhan, L.; Li, W. Effect of pre-strain on creep aging behavior of 2524 aluminum alloy. *J. Alloys Compd.* **2017**, *691*, 564–571. [CrossRef]

3. Yang, Y.; Zhan, L.; Shen, R.; Yin, X.; Li, X.; Li, W.; Huang, M.; He, D. Effect of pre-deformation on creep age forming of 2219 aluminum alloy: Experimental and constitutive modelling. *Mater. Sci. Eng. A* **2017**, *683*, 227–235. [CrossRef]

4. Lei, C.; Yang, H.; Li, H.; Shi, N.; Zhan, L. Dependences of microstructures and properties on initial tempers of creep aged 7050 aluminum alloy. *J. Mater. Process. Technol.* **2017**, *239*, 125–132. [CrossRef]

5. Xu, Y.; Zhan, L.; Ma, Z.; Huang, M.; Wang, K.; Sun, Z. Effect of heating rate on creep aging behavior of Al-Cu-Mg alloy. *Mater. Sci. Eng. A* **2017**, *688*, 488–497. [CrossRef]

6. Ho, K.C.; Lin, J.; Dean, T.A. Modelling of springback in creep forming thick aluminum sheets. *Int. J. Plast.* **2004**, *20*, 733–751. [CrossRef]

7. Zhan, L.; Lin, J. , Dean, T.A.; Huang, M. Experimental studies and constitutive modelling of the hardening of aluminium alloy 7055 under creep age forming conditions. *Int. J. Mech. Sci.* **2011**, *53*, 595–605. [CrossRef]

8. Xu, F.; Zhang, J.; Deng, Y.; Zhang, X. Precipitation orientation effect of 2124 aluminum alloy in creep aging. *Trans. Nonferrous Met. Soc. China* **2014**, *24*, 2067–2071. [CrossRef]

9. Yang, Y.; Zhan, L.; Li, J. Constitutive modeling and springback simulation for 2524 aluminum alloy in creep age forming. *Trans. Nonferrous Met. Soc. China* **2015**, *25*, 3048–3055. [CrossRef]

10. Chen, Y.; Pan, S.; Zhou, M.; Yi, D.; Xu, D.; Xu, Y. Effects of inclusions, grain boundaries and grain orientations on the fatigue crack initiation and propagation behavior of 2524-T3 Al alloy. *Mater. Sci. Eng. A* **2013**, *580*, 150–158. [CrossRef]

11. Siddiqui, R.A.; Abdul-Wahab, S.A.; Pervez, T. Effect of aging time and aging temperature on fatigue and fracture behavior of 6063 aluminum alloy under seawater influence. *Mater. Des.* **2008**, *29*, 70–79. [CrossRef]

12. Liu, Y.; Liu, Z.; Li, Y.; Xia, Q.; Zhou, J. Enhanced fatigue crack propagation resistance of an Al-Cu-Mg alloy by artificial aging. *Mater. Sci. Eng. A* **2008**, *527*, 4070–4075.

13. Chen, X.; Liu, Z.; Lin, M.; Ning, A.; Zeng, S. Enhanced Fatigue Crack Propagation Resistance in an Al-Zn-Mg-Cu Alloy by Retrogression and Reaging Treatment. *J. Mater. Eng. Perform.* **2012**, *21*, 2345–2353. [CrossRef]

14. Zhou, Z.; Zhang, J.; Deng, Y. Creep forming heat treatment technology of Al-Cu-Mg alloy. *Chin. J. Nonferrous Met.* **2016**, *26*, 1607–1614. (In Chinese)

15. Chen, Y.; Deng, Y.; Wan, L.; Jin, K.; Xiao, Z. Microstructures and Properties of 7050 Aluminum Alloy Sheet During Creep Aging. *J. Mater. Eng.* **2012**, *2*, 71–76. (In Chinese)

16. Quan, L.; Zhao, G.; Gao, S.; Muddle, B.C. Effect of pre-stretching on microstructure of aged 2524 aluminium alloy. *Trans. Nonferrous Met. Soc. China* **2011**, *21*, 1957–1962. [CrossRef]

17. Marceau, R.; Sha, G.; Ferragut, R.; Dupasquier, A.; Ringer, S. Solute clustering in Al-Cu-Mg alloys during the early stages of elevated temperature ageing. *Acta Mater.* **2010**, *58*, 4923–4939. [CrossRef]

18. Starke, E.A.; Sraley, J.T. Application of modern aluminum alloys to aircraft. *Prog. Aerosp. Sci.* **1996**, *32*, 131–172. [CrossRef]

19. Huda, Z.; Zaharinie, T.; Min, G. Temperature effects on material behavior of aerospace aluminum alloys for subsonic and supersonic aircraft. *J. Aerosp. Eng.* **2010**, *23*, 124–128. [CrossRef]

20. Zheng, Z.; Cai, B.; Zhai, T.; Li, S. The behavior of fatigue crack initiation and propagation in AA2524-T34 alloy. *Mater. Sci. Eng. A* **2011**, *528*, 2017–2022. [CrossRef]

21. Li, W.; Zhan, L.; Liu, L.; Xu, Y. The effect of creep aging on the fatigue fracture behavior of 2524 aluminum alloy. *Metals* **2016**, *6*, 215. [CrossRef]

22. Xu, Y.; Zhan, L.; Xu, L.; Huang, M. Experimental research on creep aging behavior of Al-Cu-Mg alloy with tensile and compressive stresses. *Mater. Sci. Eng. A* **2017**, *682*, 54–62. [CrossRef]

23. Bray, G.H.; Glazov, M.; Rioj, R.J.; Lib, D.; Gangloffb, R.P. Effect of artificial aging on the fatigue crack propagation resistance of 2000 series aluminum alloys. *Int. J. Fatigue* **2001**, *23*, 265–276. [CrossRef]

24. Liu, Z.; Chen, J.; Wang, S.; Yuan, D.; Yin, M.; Wu, C. The structure and the properties of S-phase in AlCuMg alloys. *Acta Mater.* **2011**, *59*, 7396–7405. [CrossRef]

25. Ratchev, P.; Verlinden, B.; Smet, P.; Houtte, P. Precipitation hardening of an Al-4.2 wt % Mg-0.6 wt % Cu alloy. *Acta Mater.* **1998**, *46*, 3523–3533. [CrossRef]

26. Sha, G.; Marceau, R.K.W.; Gao, X.; Muddle, B.C.; Ringer, S.P. Nanostructure of aluminium alloy 2024: Segregation, clustering and precipitation processes. *Acta Mater.* **2011**, *59*, 1659–1670. [CrossRef]

27. Kamp, N.; Gao, N.; Starink, M.J.; Sinclair, I. Influence of grain structure and slip planarity on fatigue crack growth in low alloying artificially aged 2xxx aluminium alloys. *Int. J. Fatigue* **2007**, *29*, 869–878. [CrossRef]

28. Liu, M.; Liu, Z.; Bai, S.; Xia, P.; Ying, P.; Zeng, S. Solute cluster size effect on the fatigue crack propagation resistance of an underaged Al-Cu-Mg alloy. *Int. J. Fatigue* **2016**, *84*, 104–112. [CrossRef]

metals

MDPI

Article

Experimental Investigations of the In-Die Quenching Efficiency and Die Surface Temperature of Hot Stamping Aluminium Alloys

Kailun Zheng [1], Junyi Lee [1], Wenchao Xiao [2], Baoyu Wang [2,*] and Jianguo Lin [1]

[1] Department of Mechanical Engineering, Imperial College London, Exhibition Road, London SW7 2AZ, UK; k.zheng13@imperial.ac.uk (K.Z.); junyi.lee108@imperial.ac.uk (J.L.); jianguo.lin@imperial.ac.uk (J.L.)

[2] School of Mechanical Engineering, University of Science and Technology Beijing, Beijing 100083, China; xwcxiaowenchao@163.com

* Correspondence: bywang@ustb.edu.cn; Tel.: +86-010-8237-5671

Received: 28 February 2018; Accepted: 27 March 2018; Published: 2 April 2018

Abstract: The in-die quenching is a key stage in the hot stamping volume production chain which determines the post-formed strength of lightweight alloy components, tool life, and hot stamping productivity. In this paper, the performance of in-die quenching, reflected by the quenching efficiency (the time of work-piece held within stamping dies) and die surface temperature during the simulated hot stamping process of AA6082, was experimentally and analytically investigated. A range of in-die quenching experiments were performed for different initial work-piece and die temperatures, quenching pressures, work-piece thickness, and die clearances, under hot stamping conditions. In addition, a one-dimensional (1D) closed-form heat transfer model was used to calculate the die surface temperature evolution that is difficult to obtain during practical manufacture situations. The results have shown that the in-die quenching efficiency can be significantly increased by decreasing the initial work-piece and die temperatures. Die clearances are required to be designed precisely to obtain sufficiently high quenching rates and satisfying post-formed strength for hot-stamped panel components. This study systematically considered an extensive variety of influencing factors on the in-die quenching performance, which can provide practical guides for stamping tool designers and manufacture systems for hot-stamping volume production.

Keywords: in-die quenching; hot stamping; heat transfer; aluminium alloy

---

1. Introduction

The increasing concern of air pollution and stringent legislation of greenhouse gas emissions in the transportation industry has driven the automotive and aircraft industries to use lightweight materials, such as aluminium, magnesium alloys, and composites [1,2]. Hot Form and Quench (HFQ®) is a novel and leading hot stamping technique to manufacture complex-shaped panel components of high-strength aluminium alloys [3]. During the HFQ® process, the raw aluminium alloy blank experiences solution heat treatment (SHT) to dissolve the original precipitates and dissolvable inclusions to obtain a ductile microstructure. Then, the heat treated blank is transferred to the tools and hot stamped into designed geometries. Similar to hot stamped boron steels, the hot stamped aluminium alloy components, especially for heat-treatable aluminium alloys, are required to be cold die-quenched to lower temperatures, normally less than the artificial ageing temperature. The objective of this stage is to obtain a supersaturated solid solution state, that can be artificially aged to higher post-formed strength [4], and guarantee good dimensional accuracy. Therefore, the in-die quenching is an essential and critical stage of hot-stamping volume production. The performance of the in-die quenching stage determines the post-formed strength of formed components [5], stamping tool life [6], and productivity [7], which needs to be thoroughly investigated and precisely controlled.

Heat-treatable aluminium alloys are strengthened through precipitation hardening. Hence, the post-formed strength of hot stamped aluminium alloys components is normally achieved by artificial ageing, which requires the alloy pre-quenched at a sufficiently high cooling-rate after solution heat treatment. In hot stamping aluminium alloys, such a cooling rate is obtained by in-die quenching similar to hot stamping boron steel. Until now, extensive studies have been performed on the cold-die quenching of press-hardened boron steel [8]. Intrinsically, the quenching rate is, in part, determined by the interfacial heat transfer coefficient (IHTC) between the stamping dies and formed alloys. The IHTC varies with various factors, such as the contact material pair [9], die temperature [10], contact pressure [11], surface roughness [12], lubrication [13], oxidation [14], and clearance [15], which results in the varied in-die quenching time. However, the research of in-die quenching of hot stamping aluminium alloys is still limited. Ying et al. [16] thoroughly investigated the heat transfer mechanism of HFQ® forming AA7075. The IHTC increases with increasing contact pressure within a certain range, 30–80 MPa. In addition, surface roughness and lubrication also affect the IHTC significantly. A surface roughness between 0.5 and 1 μm, and a lubricant with larger heat conductivity enhances the interfacial heat transfer. Similar findings were also obtained by Xiao et al. [17] for hot stamping AA7075. Liu et al. [18] developed a novel testing facility to measure the temperature evolutions of blank and die materials using the Gleeble 3800 thermal-mechanical simulator. The IHTCs with different tool materials and lubrication at different contact pressures were obtained using an inverse finite element (FE) methodology. The studies described previously concentrated on the determination of the IHTC values as functions of contact pressures, contact material pair and lubrication, which is believed to be useful for FE simulations.

During the volume production of hot stamping aluminium alloys, the stamping dies are subjected to severe interface conditions, such as high temperature and relatively high stress. After certain cycles of hot stamping operations, the increasing die-surface temperatures results in cooling water being required to cool the material [8]. The increased die surface temperature may induce drawbacks, including extended in-die quenching time, insufficient quenching rate, and shortened tool life due to thermal fatigue. Subsequently, the manufacturing productivity is reduced, and the strengths of hot-stamped components deteriorate. However, systematic investigations on the die temperature evolution and resulted in-die quenching efficiency under different process parameters of hot stamping aluminium alloys are still lacking.

The die surface temperature is difficult to measure precisely during hot stamping using conventional experimental methods since the die surface is always in contact with the hot work-piece. Additionally, the in-die quenching efficiency and effective quenching rate vary with several process variables, such as different initial alloy temperatures, work-piece thickness, die clearance, and contact pressure. In this study, a series of in-die quenching experiments with a variety of process parameters mentioned above were performed to evaluate the in-die quenching efficiency. In addition, a one-dimensional (1D) closed-form heat transfer model was utilised to calculate the die surface temperature evolution [19] using the experimentally-determined temperature evolutions. This is the first study that systematically investigates the in-die quenching performance from the viewpoint of hot stamping volume production. The experimental results are used to guide tooling designs and manufacturing systems of hot stamping aluminium alloys.

2. Experimentation

2.1. Material and Specimen Design

The raw material used was AA6082 in T6 condition supplied by Southwest Aluminum (Group) Co., Ltd., Chongqing, China. AA6082 is a high-strength aluminium alloy with good corrosion properties and preferably used as a material in automobile body structures. Two different thickness, 2 mm and 3 mm, were selected. The chemical composition of AA6082 is given in Table 1.

Table 1. Main chemical composition of AA6082.

Element wt %	Mg	Si	Mn	Fe	Cu	Zn	Ti	Cr	Remainder
Min.	0.6	0.7	0.4	-	-	-	-	-	Al
Max.	1.2	1.3	1	0.5	0.1	0.2	0.1	0.25	

Figure 1 shows the dimensions of the circular specimen used for the in-die quenching test. Circular specimens with diameters of 100 and 150 mm were used. Four holes with a diameter of 1 mm, distributed orthogonally, were machined using electric discharging machining (EDM) through the middle of the thickness to different depths from the edge to the centre of the specimen. Thermo-couples were attached to the bottom surface of the drilled holes to allow for the precise measurements of the specimen temperature and to avoid edge effects.

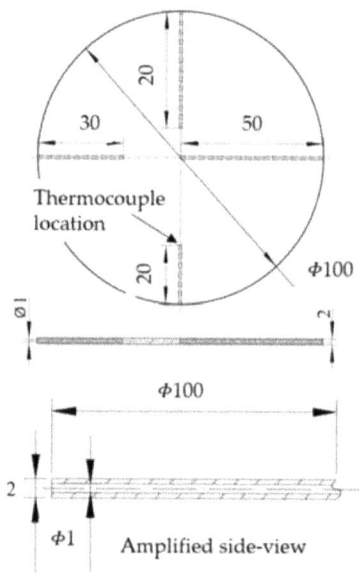

Figure 1. Thermocouple layout in the blank to measure the blank (100 mm diameter and 2 mm thick) temperature at different positions (all dimensions are in mm).

2.2. Experimental Setup

Figure 7a shows the dimensions and positions of the in-die quenching tools used to simulate the in-die quenching stage in the practical hot stamping process of aluminium alloys. In this set-up, cylindrical dies with a diameter of 112 mm were used. The tool was made of G3500 cast iron. Three thermocouples were welded inside the lower die at 1.5, 3, and 4 mm from the die surface. Figure 7b shows the complete set-up of the in-die quenching test. The die quenching rig was mounted on a 250 kN hydraulic press purchased from E.S.H. Testing Limited (Birmingham, UK). The base of this rig was supported by a gas-cushion system to provide different counter forces to simulate the in-die quenching process. The temperature evolutions of both dies and test-piece were recorded using a TC-08 Pico data logger (Pico Technology, Cambridgeshire, UK). The dies were heated to different temperatures using the heating bands attached on the periphery of the upper and lower dies in this study to simulate the temperature rise of stamping dies during the volume production of hot stamping. A Proportional Integral Derivative (PID) control system was used to control the die temperature. The upper and bottom dies were assumed to be at the same temperature due to the symmetrical geometry of the setup. Additionally, due to the clearance between the forming dies and the localised thinned areas on the formed part, there might exist some gaps between the dies and formed part, and full contact cannot be guaranteed. To simulate the clearance effect on quenching, single- and double-sided clearances were implemented into the in-die quenching rig used in previously-published research [17], as shown in Figure 7c. The clearances, either single-sided or double-sided, were obtained by inserting clearance blocks (stainless steel foils) between the upper and lower dies. The different magnitudes of clearance were achieved by altering the number of foils.

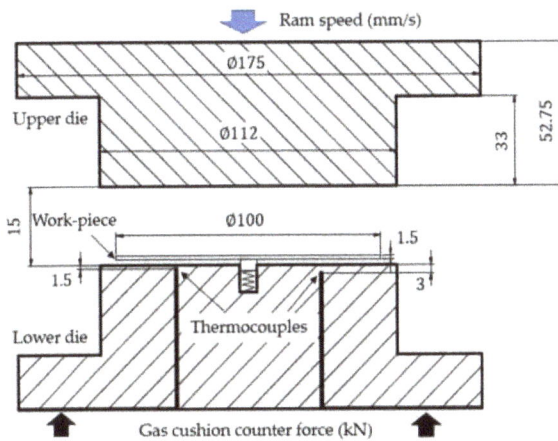

Figure 2. *Cont.*

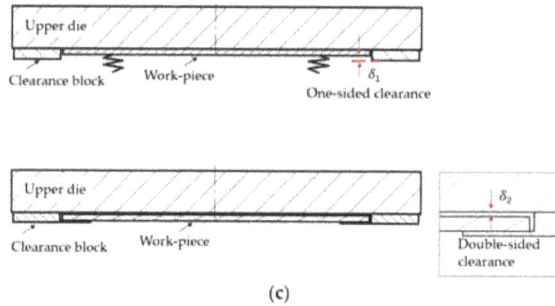

Figure 2. Experimental set-up of in-die quenching: (**a**) test rig dimensions (all dimensions are in mm); (**b**) in-die quenching test rig; and (**c**) schematic of die clearance designs.

2.3. Test Programme

Figure 3a illustrates the temperature profiles used to assess the in-die quenching under the HFQ® technology forming conditions (solid line). In order to achieve different temperatures before in-die quenching, a fast intermediate-cooling stage is preferably used, enabling the preservation of the obtained optimal microstructure during solution heat treatment (SHT) as disclosed in the patent WO2015136299. Due to difficulties in controlling the temperature during the experiments, a modified simple temperature profile, as indicated by the dashed line, was used in this study to replace the conventional HFQ® technology process in Figure 3b. The effects of material properties on the thermal response are assumed to be negligible. In this process, the as-received work-piece was heated to different temperatures, and soaked for 1 min to achieve a uniform temperature distribution. Then, the work-piece was transferred to the pins located on the lower die and air cooled to target temperatures. Subsequently, the ram of the press was activated to press the upper die at a speed of 300 mm/s to close the 15 mm gap between the upper and lower dies. The in-die quenching force was exerted by the counterforce of gas cushion system below the whole rig. The in-die quenching stage was completed, and the upper and lower dies were separated once the work-piece was quenched to designed temperatures that are lower than the artificial ageing temperature of AA6082 (190 °C). To quantify the in-die quenching efficiency, a term, t_{quench}, equaling to the time of quenching the work-piece from an initial temperature $T_{initial}$ to a target quenched temperature T_{quench}, is used. These values need to be determined experimentally. Before each test, gas springs were charged to different pressures to provide different in-die quenching forces. The charged pressures were able to achieve different contact pressures as summarised in Table 2. In addition, the quenching dies were heated to different temperatures. During quenching, the temperatures of both the work-piece and the dies were continually recorded. Table 2 summarises the conditions of the in-die quenching tests.

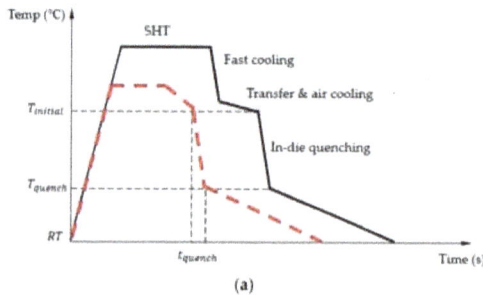

Figure 3. *Cont.*

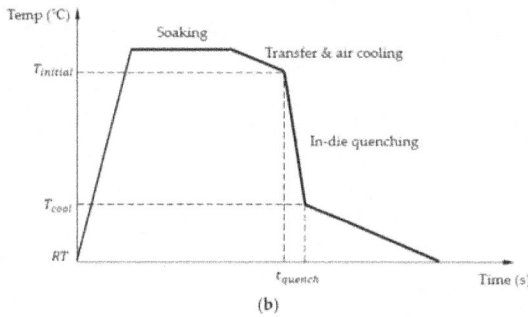

(b)

Figure 3. Temperature profile of work-piece during in-die quenching process: (**a**) HFQ® technology for lower temperatures (solid line) and alternative process (dash line); and (**b**) conventional forming using HFQ® technology.

Table 2. Test matrix of in-die quenching efficiency test.

Work-Piece Temperature (°C)	Work-Piece Thickness (mm)	Die Temperature (°C)	Contact Pressure (MPa)
250, 350, 450, 525	2, 3	25, 50, 75, 100	0.03, 1, 10

Figure 3b shows the temperature profile of the conventional HFQ® process, which was used to investigate effects of die clearances on the in-die quenching performance. Initially, the work-piece was heated to 525 °C, and soaked for 2 min. Then, it was transferred to the quenching dies and in-die quenched with different clearances. The work-piece thickness was 2 mm. The values of clearance used were 0.1, 0.2, 0.5, and 1 mm for both single-sided and double-sided clearances. Before each test, a different number of steel foils were inserted. The accuracy of the clearance was controlled to within 0.02 mm.

3. Die Surface Temperature Calculation

Since the work-piece material flows at the interface between die surfaces, the die surface temperature evolution during hot stamping is crucial, which may affect the friction of tooling and lubrication, such as surface coating. In this study, a one-dimensional (1D) closed-form method was used to calculate the die surface temperatures under different process variables. The temperature evolution of an object can be calculated by the energy conservation equation, given by Equation (1). Due to the absence of mass transfer, thermal radiation and additional heat generated by plastic straining, these effects were neglected during the in-die quenching stage:

$$\frac{\partial T}{\partial t} = \frac{k}{\rho c_{\mathrm{p}}} \nabla^2 T \tag{1}$$

where T is the temperature, t is time, ρ is the density, k is the thermal conductivity, and c_{p} is the thermal capacity. The heat transfer between the cold dies and the hot work-piece is simplified to a one-dimensional heat transfer condition, with heat flow along the thickness of the work-piece. The heat transfer in the radial direction was ignored because the central position of thermocouples results in the convection by the surrounding air being negligible. Based on this assumption, the differential equation in Equation (1) can be rewritten as below:

$$\frac{\partial T}{\partial t} = \alpha \frac{\partial T^2}{\partial x^2} \tag{2}$$

where *x* represents the position and:

$$\alpha = \frac{k}{\rho c_p} \tag{3}$$

To solve the partial differential equation, Equation (2), a finite difference method, the backward time-centred space difference scheme, described by Equation (4), was used:

$$\frac{T_x^t - T_x^{t-\Delta t}}{\Delta t} = \alpha \frac{T_{x-\Delta x}^t - 2T_x^t + T_{x+\Delta x}^t}{\Delta x^2} + O(\Delta t) + O\left(\Delta t^2\right) \tag{4}$$

where Δt is the time increment, Δx is the distance increment within work-piece or die. O is the truncation error. After neglecting the higher-order truncation errors, Equation (4) can be recast as follows:

$$-\frac{\alpha \Delta t}{\Delta x^2} T_{x-\Delta x}^t + \left(1 + \frac{2\alpha \Delta t}{\Delta x^2}\right) T_x^t - \frac{\alpha \Delta t}{\Delta x^2} T_{x+\Delta x}^t = T_x^{t-\Delta t} \tag{5}$$

Then, a matrix of the temperature evolution is established from Equation (5):

$$MT^t = \begin{vmatrix} b_1 & c_1 & 0 & 0 & \cdots & 0 & 0 & 0 & 0 \\ a_2 & b_2 & c_2 & 0 & \cdots & 0 & 0 & 0 & 0 \\ 0 & a_3 & b_3 & c_3 & \cdots & 0 & 0 & 0 & 0 \\ * & * & * & * & * & * & * & * & * & * \\ 0 & 0 & 0 & 0 & \cdots & a_{N-2} & b_{N-2} & c_{N-2} & 0 \\ 0 & 0 & 0 & 0 & \cdots & 0 & a_{N-1} & b_{N-1} & c_{N-1} \\ 0 & 0 & 0 & 0 & \cdots & 0 & 0 & a_N & b_N \end{vmatrix} \begin{Vmatrix} T_1^t \\ T_2^t \\ T_3^t \\ * \\ T_{N-2}^t \\ T_{N-1}^t \\ T_N^t \end{Vmatrix} = \begin{vmatrix} T_1^{t-\Delta t} \\ T_2^{t-\Delta t} \\ T_3^{t-\Delta t} \\ * \\ T_{N-2}^{t-\Delta t} \\ T_{N-1}^{t-\Delta t} \\ T_N^{t-\Delta t} \end{vmatrix} \tag{6}$$

where the subscripts denote the node number, as shown in Figure 4. When $x = 0$, $i = 0$. The coefficients *a*, *b*, and *c* are defined as follows:

$$a = -\frac{\alpha \Delta t}{\Delta x^2}$$
$$b = 1 + \frac{2\alpha \Delta t}{\Delta x^2} \tag{7}$$
$$c = 1 + \frac{2\alpha \Delta t}{\Delta x^2}$$

The temperature at each time step and location within the die can be calculated when the initial condition, T^0 (initial die temperature), T_1^t and T_N^t are known. The initial die temperature was assumed to be uniform, i.e., 25 °C, 50 °C, 75 °C, and 100 °C. $T_{\text{Die}}^{\text{Surf-N}}$ is the temperature of the die recorded from the 4 mm thermocouple. The die surface temperature, $T_{\text{Die}}^{\text{Surf}}$, which is the other boundary condition, was determined by first setting an initial guess of this value and then calculated by using the least-squares method to minimise the difference between calculated and measured temperatures at the other locations. The error function is given as:

$$error_t = \sum_{i=1}^{N} ((T_{\text{Die-e}}(i) - T_{\text{Die\_c}}(i)))^2 \tag{8}$$

where $T_{\text{Die\_e}}(i)$ represents the experimentally-measured temperature of die location and $T_{\text{Die\_c}}(i)$ represents the calculated temperature using the surface temperature that is set during the optimisation process. An optimisation algorithm was used to determine the die surface temperature so that the minimum error is obtained.

This procedure is then repeated for every time step. Lastly, the heat flux, $\dot{q}$, across the surface of the work-piece and die was calculated using the gradients at these surfaces, which is given by the following equation based on the finite difference scheme. This approach is similar to the one performed in [19]:

$$\dot{q} = k\frac{\partial T}{\partial x} = k\frac{T_{\text{Surf}} - T_{(\text{Surf-1})}}{x_{\text{Surf}} - x_{(\text{Surf-1})}} \tag{9}$$

where the subscript surf denotes the point at the surface.

The effective heat transfer coefficient, h, is then calculated using the average heat flux of the surface of the die, $\dot{q}_{Die}$, and work-piece, $\dot{q}_{WP}$:

$$h = \frac{T_{Die}^{Surf} - T_{WP}^{Surf}}{0.5\left(\dot{q}_{WP} + \dot{q}_{Die}\right)} \tag{10}$$

The work-piece temperature is calculated straightforwardly using Equation (6), and setting the boundary conditions to the appropriate values. The temperature at the surface of the work-piece was set to be that of the temperature measured by the thermocouples. The work-piece is assumed to be symmetrical, so the thermal gradient at the other end, which is the line of symmetry, is set to be zero.

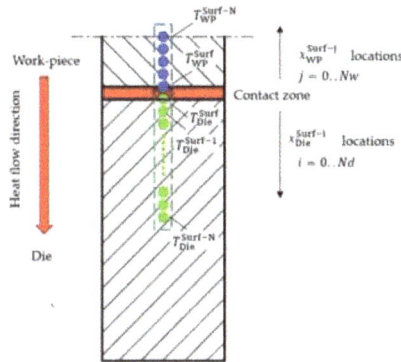

Figure 4. Simplified 1D heat transfer finite difference method for the work-piece and die with different initial temperatures.

4. Results and Discussion

4.1. Validation of 1D Closed Form Calculation Method

The 1D closed form calculations were verified by both the experimental results and finite element simulations. Figure 5 shows the comparison between experimental results of the in-die quenching test and the calculated results for the work-piece and die temperatures. The initial temperatures of work-piece and die were 525 °C and 25 °C, respectively. The interface contact pressure was 10 MPa. For the die, temperatures at three locations, which are 1.5 mm, 3 mm, and 4 mm from the die surface, were measured (solid symbols) and compared with the computational results (solid lines). As shown in Figure 5, in general, the computed results using the 1D closed-form model exhibit a good agreement with the experimentally-measured results. The characteristics of temperature evolutions of both the work-piece and die can be predicted. Initially, the temperature of the work-piece dropped drastically in the first 1.5 s, and the significant heat transferred from the work-piece to the die resulted in a temperature increase on the areas approaching die surface. Then, the work-piece temperature decreased gradually due to the small temperature gradient between the work-piece and die surface. Less heat was transferred to the die, subsequently, and the heat was further transferred from the die surface to the inner portions of the die, which enables the temperature near the die surface (1.5 mm) to decrease, and the temperature away from the die surface (4 mm) to increase gradually. In addition, the interface heat transfer coefficient was calculated and implemented into the finite element (FE) simulation model, Deform 2D, to predict the temperature history of work-piece and die surface temperature evolution, as performed in a previous study [17]. Good agreement is also observed for the work-piece between experimental and computed results, as shown in Figure 5, which validates the calculation of the 1D closed-form model.

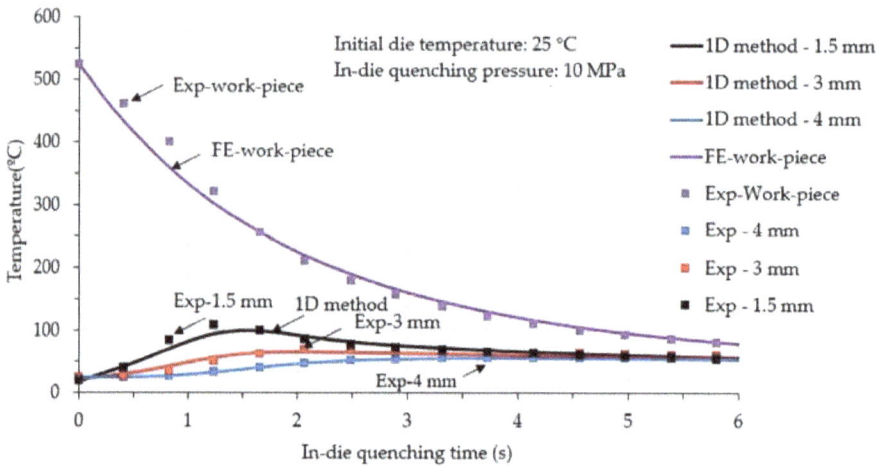

Figure 5. Validation of 1D closed-form method and FE simulation, where solid symbols represent experimentations, and solid lines represent calculation and computational results. The blank thickness was 3 mm.

4.2. In-Die Quenching Efficiency

During the continuous volume production of hot stamping aluminium alloy components, the in-die quenching time (time of formed component held by stamping dies) is an important parameter that needs to be designed. This parameter determines the manufacturing efficiency and productivity. For the HFQ® forming of aluminium alloy parts, after hot stamping, the stamping dies are separated when the part temperature is quenched to a temperature that is lower than the artificial ageing temperature. Figure 6 shows the required time of cooling work-piece, t_{quench}, to different temperatures, T_{quench}, with different initial work-piece and die temperatures. As seen in Figure 6a, the in-die quenching time increases significantly with decreasing T_{quench}, which is the work-piece temperature after in-die quenching. Furthermore, the time decreases with decreasing initial work-piece temperature. However, the reduction of this time is not obvious, unless T_{quench} is sufficiently low, such as 40 °C. Figure 6b shows variations of in-die quenching time of quenching the work-piece to different temperatures at different initial die temperatures, which the simulation of the effects of increased die temperature after a certain number of hot stamping cycles during mass production. To cool the work-piece to 80 °C, the in-die quenching time of the 75 °C die was severely increased to above 25 s compared to the 5 s of the die at 25 °C (room temperature). Therefore, the die temperature should be below 75 °C for the hot stamping of aluminium alloys.

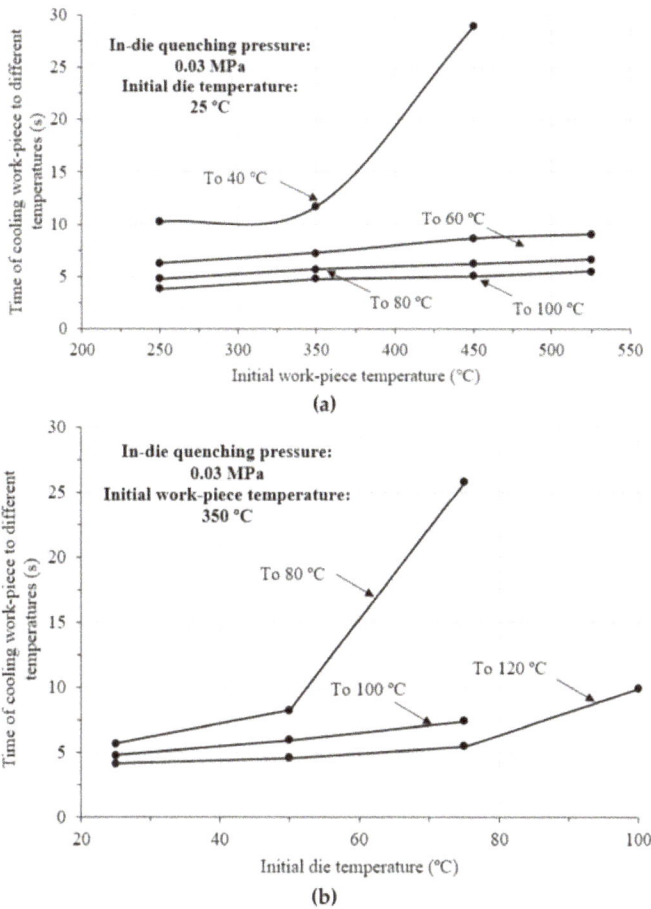

Figure 6. Effects of initial blank and die temperatures on the in-die quenching time of cooling work-piece to different temperatures at an in-die quenching pressure 0.03 MPa. The work-piece thickness was 2 mm. (**a**) Different initial work-piece temperatures; and (**b**) different initial die temperature.

4.3. Evolutions of Die Surface Temperature

Figure 7 shows the calculated die temperature evolutions under different influencing factors. Figure 7a shows variations of die surface temperature with different initial work-piece temperatures. The initial die temperature was 25 °C, and the contact pressure was 0.03 MPa. With the decreasing initial work-piece temperature of in-die quenching, the die surface temperature decreases significantly, which indicates that forming at relatively lower temperatures, unlike forming using the conventional HFQ® technology, can cause the severe high-temperature condition of the die surface and contribute to reducing thermal fatigue and potentially extending the tool life. Additionally, a relatively lower interface temperature also results in lower friction, which is believed to be beneficial for flanged material flow. Figure 7b shows the contact pressure effect on the die surface temperature evolutions. At a high contact pressure of 10 MPa, the maximum die surface temperature increased greatly. For instance, for the conventional HFQ® process with a die temperature at room temperature. The maximum die surface temperature increased to 150 °C compared to the 85 °C of 0.03 MPa. Therefore, the blank-holding pressure is normally controlled not to be too high, to avoid the increase of the interface temperature and friction coefficient correspondingly. In addition, the profile of the die surface temperature evolution also varies with contact pressure. For high temperatures, the heat within the work-piece is transferred quickly to the die due to the high contact pressure, because of the greater deformation of surface asperities. This situation results in the die surface temperature increasing significantly at the early stage of in-die quenching. Then, the work-piece and die surface temperatures approached a steady state, and the heat transferred from the die surface to the inside regions of the die are now more than the input heat from the work-piece, resulting in a decrease of the die surface. In comparison, at a low contact pressure of 0.03 MPa, the temperature of the die surface increased gradually due to the slower heat input from the work-piece. Correspondingly, the quenching rate of the work-piece was lower. Therefore, using a die material with high thermal conductivity is beneficial to the in-die quenching efficiency, combined with using a contact pressure, the heat at the die surface is able to be immediately transferred into the die, resulting in high quenching rates and less die surface temperature effects. Figure 7c shows the die surface temperature evolutions of different initial temperatures. The contact pressure used was 8 MPa, which is similar to Figure 7b. The evolution trends are similar for different temperatures, which can be regarded as an offset with the magnitude of the initial die temperature.

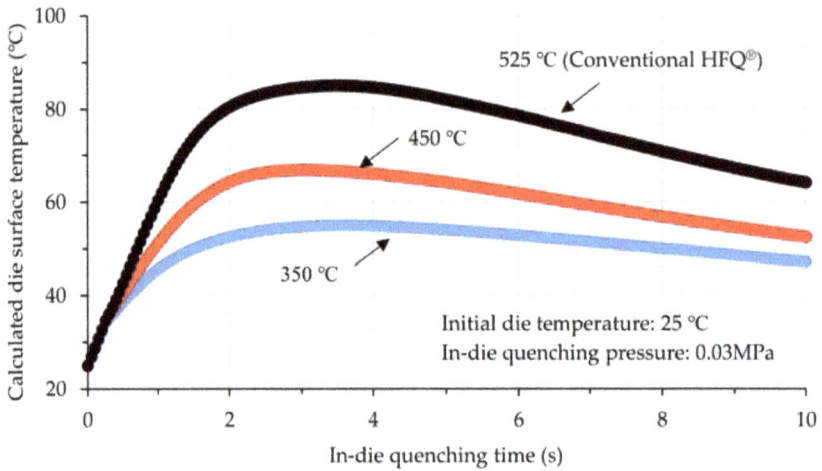

(a)

Figure 7. *Cont.*

(b)

(c)

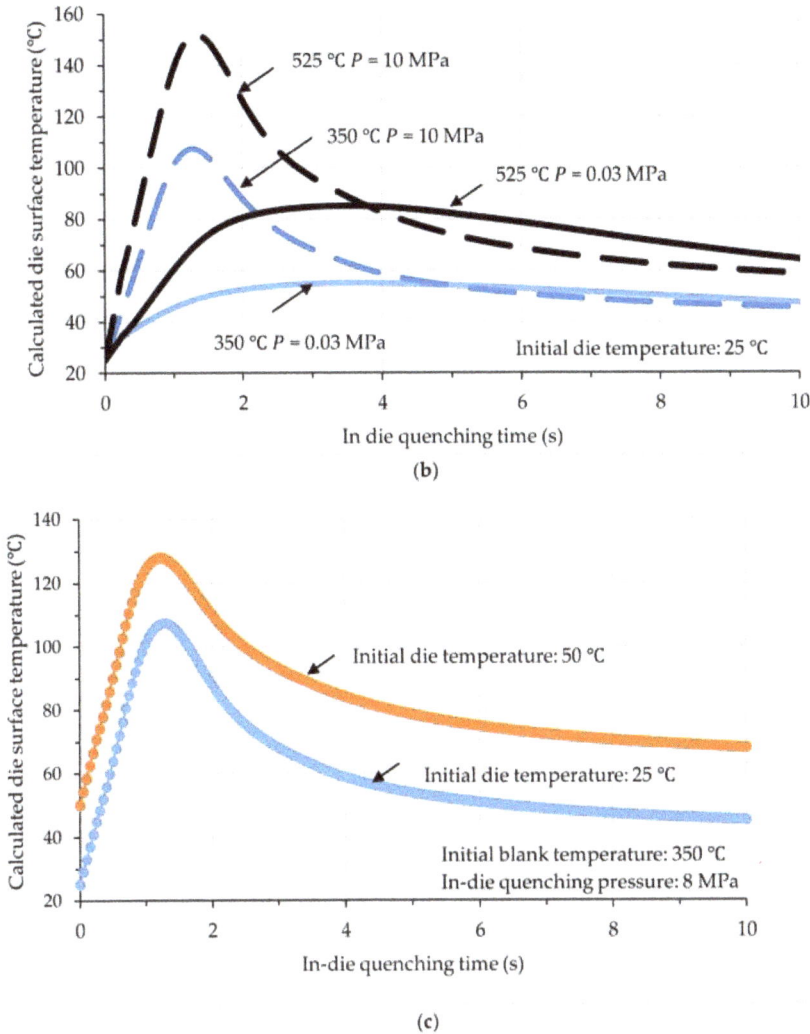

Figure 7. Calculated die surface temperatures at different process conditions during in-die quenching. The work-piece thickness was 3 mm. (**a**) Different initial work-piece temperature; (**b**) different contact pressure; and (**c**) different initial die temperatures.

4.4. Effects of Contact Pressure

The interfacial heat transfer coefficient is dependent on the interface contact pressure, which determines the in-die quenching efficiency significantly. Figure 8 shows effects of contact pressure on the in-die quenching time when T_{quench} equals to 100 °C with different initial work-piece temperatures. For a given initial work-piece temperature, by increasing the contact pressure to 4 Mpa, the required t_{quench} is significantly reduced. However, the time cannot be further reduced by increasing the contact pressure. In addition, t_{quench} also increases with increasing initial work-piece temperature at a given contact pressure. For a contact pressure of 4 MPa, the t_{quench} with an initial work-piece temperature of 525 °C increases to 4.7 s, which is an approximately 56.7% increase compared to the 3 s using an initial work-piece temperature of 350 °C. This observation suggests that reducing the initial forming temperature of the HFQ® process contributes to an improved in-die quenching efficiency. The effects on the maximum die surface temperatures are shown in Figure 7b. For an initial work-piece temperature of 525 °C, the maximum die surface of 10 MPa was higher than that of 0.03 MPa. The reason for this observation is that the interfacial heat transfer increases with contact pressure, resulting in more heat being transferred to the die material (regions approaching the surface) and had no time to be further be transferred within the die material during the early stage of quenching. Therefore, the die surface temperature increased quickly and significantly. A greater maximum die surface temperature was obtained using a higher contact pressure.

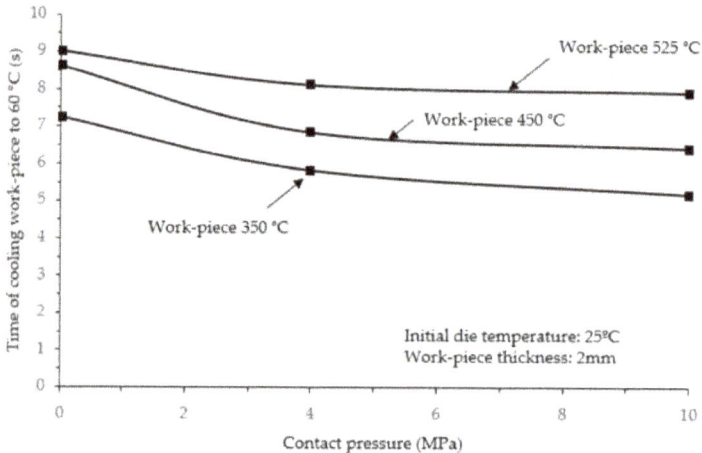

Figure 8. Effects of in-die quenching pressure on the in-die quenching efficiency. The initial die temperature was 25 °C. The work-piece thickness was 2 mm.

4.5. Effects of Die Temperature

After several hot stamping cycles, the die surface temperature will rise due to heat being transferred from the hot work-piece to the stamping dies. The increase of die surface temperature might cause the in-die quenching rate to vary and increases t_{quench} significantly. Figure 9 shows the effects of the increase in die temperature on the maximum die surface temperature and subsequent in-die quenching efficiency. As shown in Figure 9a, with the increasing die temperature, the calculated maximum die surface temperature increases severely. For an initial work-piece temperature of 450 °C, the maximum die surface temperature increases from 56.8 °C to 113.8 °C, when the temperature of die increases from room temperature to 100 °C. Such a temperature increase of the die surface results in a decrease in the interface temperature gradient, which reduces the quenching rate (Equation (1)), and the work-piece may not be able to be further cooled when the temperatures of die surface and work-piece become similar. As verified in Figure 9b, for example, the time to cool the work-piece to 80 °C increases to 25.8 s for the die at 75 °C and an initial work-piece temperature of 350 °C, which is believed to be unacceptable for continuous volume production of hot stamping aluminium alloys. In addition, it is interesting to find that the effects of initial work-piece temperature are not obvious when the die temperature is controlled below 50 °C. The reason for this observation is believed to be that, once T_{quench} equals 80 °C, the temperature gradient can still guarantee a sufficiently high cooling rate before quenching the work-piece to 80 °C. Therefore, for a practical hot stamping production line, the maximum acceptable die temperature (before water cooling dies) is dependent on the design of T_{quench} according to the age hardening performance of the specific alloy.

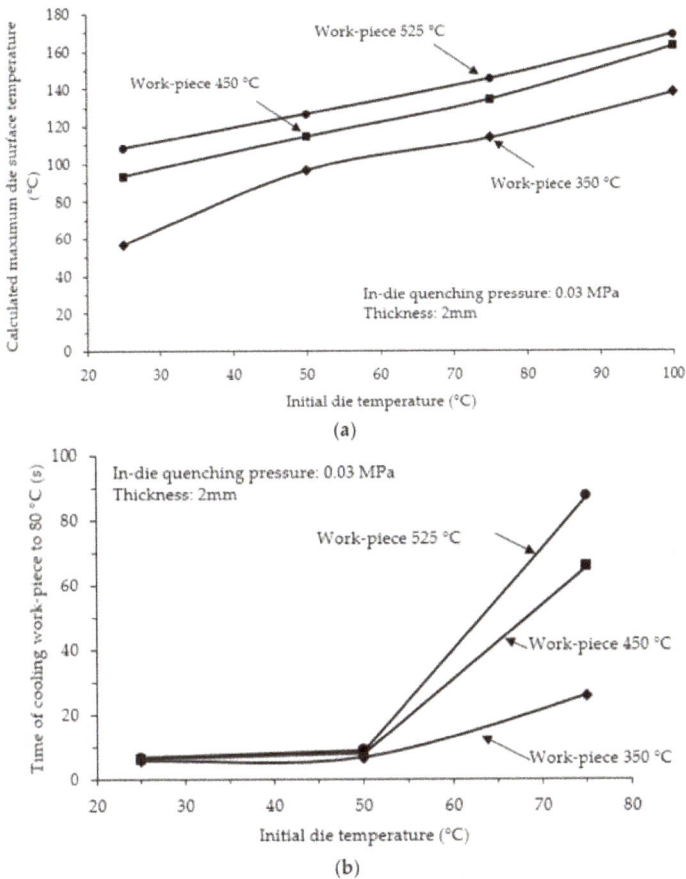

Figure 9. Effects of initial die temperature on the in-die quenching performance. The work-piece thickness was 2 mm. (**a**) Calculated maximum die surface temperature and (**b**) time of cooling work-piece to 80 °C.

4.6. Effect of Work-Piece Thickness

Figure 10 shows the effect of work-piece thickness on the in-die quenching performance. Figure 10a shows the in-die quenching efficiency, and Figure 10b shows the calculated maximum die surface temperature. As seen in Figure 10a, the thickness plays an important role in determining the in-die quenching time. For an initial work-piece temperature of 450 °C and contact pressure of 0.03 MPa, t_{quench} increases from 5 s for a thickness of 2 mm to 8.1 s for a thickness of 3 mm, which is a 62% increase of time. This indicates that a greater contact pressure is required to guarantee the efficiency of in-die quenching for thicker work-pieces. By increasing the contact pressure to 10 MPa for the 3 mm work-piece with an initial temperature of 525 °C (conventional HFQ® Technology), t_{quench} decreased from 10.5 s to 4.4 s, which is an approximately 58.1% savings in time. In addition, under a certain contact pressure, t_{quench} can be reduced by decreasing the work-piece temperature, as discussed above. For a 3 mm thick work-piece, the required t_{quench} is 4 s for an initial work-piece temperature of 350 °C, while it is 3.1 s for a 2 mm thick work-piece. In general, the thicker the work-piece, the more heat is stored within the work-piece, and the die surface temperature becomes higher because of the corresponding heat transfer, as shown in Figure 10b. For an initial die temperature of 50 °C and work-piece temperature of 525 °C, the maximum die surface temperature increased from 128.8 °C to 165.9 °C with increasing work-piece thickness from 2 mm to 3 mm at a contact pressure of 10 MPa. For a work-piece with an initial temperature of 350 °C, the maximum die surface temperature of the 3 mm workpiece at 0.03 MPa contact pressure, is similar to that of the 2 mm work-piece at 10 MPa contact pressure. On the other hand, the temperature of the 3 mm work-piece at 0.03 MPa contact pressure is higher than that of the 2 mm work-piece at 10 MPa contact pressure. This indicates that the effect of work-piece thickness should be considered in addition to the initial work-piece temperature, which is only obvious for high initial work-piece temperatures. Otherwise, the contact pressure plays a dominant role.

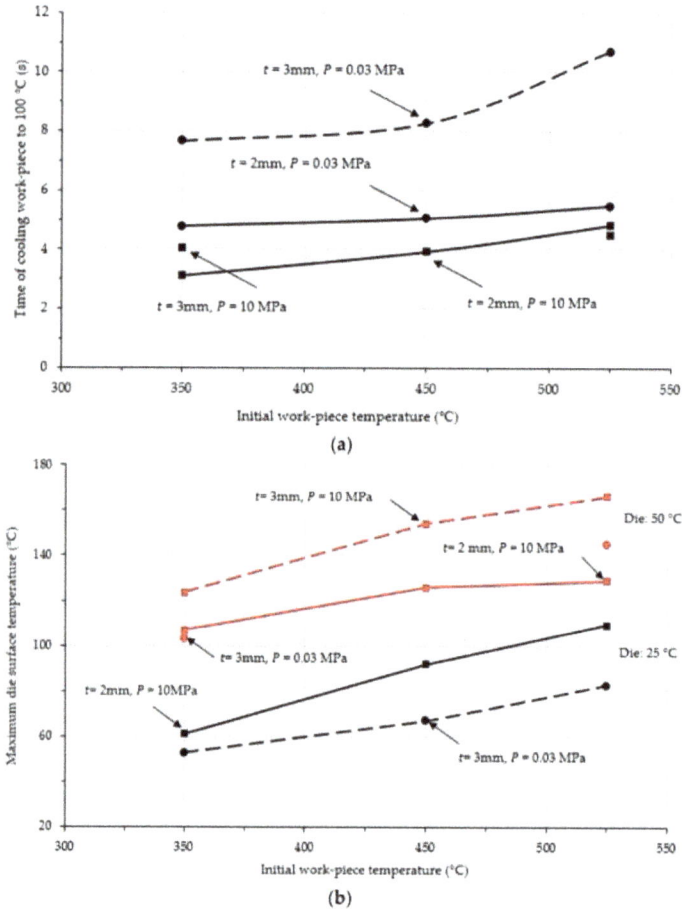

Figure 10. Effects of work-piece thickness on the in-die quenching performance, where the solid lines represent the 2 mm work-piece and the dash lines represent the 3 mm work-piece: (**a**) Time of cooling work-piece to 100 °C and (**b**) calculated maximum die surface temperature.

4.7. Effect of Die Clearance

The clearance is an important feature for hot stamping aluminium alloys due to several factors. Firstly, thinning is normally observed on a complex-shaped formed component, which results in gaps between the sheet and tools, and the contact is between the component and die is incomplete.

Secondly, there is always a clearance between the punch and the die, approximately 0.1 mm of sheet thickness, which may induce one-sided or double-sided clearances. At locations where such a clearance is present, the heat transfer between the hot work-piece to the cold dies changes to that between the hot work-piece and the air. The work-piece at these local areas is preferably cooled by the thermal conduction within work-piece, and the quenching rate might not be guaranteed, resulting in potentially lower post-formed strength, especially for high quench-sensitive aluminium alloys, such as high-strength AA7075 and AA7050, that are popular candidates for automotive use. Figure 11a shows the effects of clearance on the in-die quenching of the work-piece. The clearance used was 0.2 mm, and the initial die and work-piece temperatures are 25 °C and 450 °C, respectively. As observed in this figure, the clearance reduced the quenching rate severely. Quenching with a single-sided clearance is quicker than a double-sided clearance, as there is still a slight heat convection between the work-piece and the cold die on one side. Although quenching with double-sided clearance is similar to the scenario of air cooling, the quenching rate is still higher than that of air cooling, which is believed to be caused by the heated hot air around the work-piece being transferred to cold dies. Figure 11b shows the summarised variations t_{quench} of quenching the work-piece to 80 °C with different clearance values. The clearance, either single-sided or double-sided, severely increased t_{quench}. Apart from the initial design of the tool setup, improving the uniformity of deformation to obtain a more uniform thickness within the formed component is also critical to reducing the clearance effects.

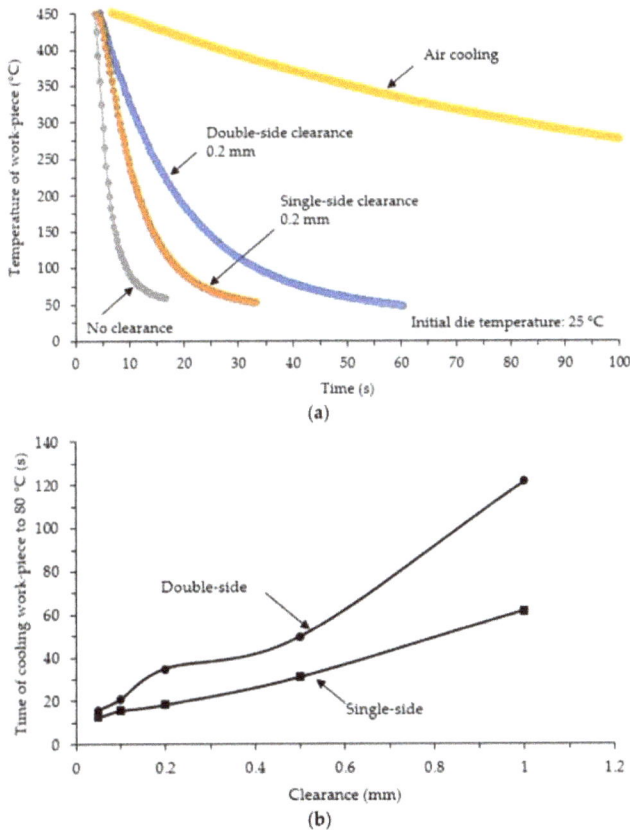

Figure 11. Effects of die clearance on (**a**) cooling of work-piece and (**b**) in-die quenching efficiency.

5. Conclusions

The presented study comprehensively investigated the key influencing factors of the in-die quenching stage of hot stamping aluminium alloys experimentally and theoretically. The performed in-die quenching tests and 1D closed-form heat transfer calculations enable the effects of initial work-piece and die temperatures, contact pressure, work-piece thickness, and clearances on the in-die quenching performance to be understood thoroughly. The following conclusions are drawn:

(1) The in-die quenching efficiency is significantly increased with decreasing initial work-piece and die temperatures and work-piece thickness;

(2) Contact pressure is the dominant parameter that determines the time of quenching and die surface temperature evolutions;

(3) The clearance of tools severely deteriorate in-die quenching performance, so precise tool design and improving the uniformity of deformation are potential solutions to address the clearance effect. In addition, the experimentally-measured and analytically-calculated data aids tooling design and thermal surface engineering treatments by providing reasonable evaluations.

Acknowledgments: The research in this paper was funded by the European Union's Horizon 2020 research and innovation programme under grant agreement No. H2020-NMBP-GV-2016 (723517) as part of the project "Low-Cost Materials Processing Technologies for Mass Production of Lightweight Vehicles (LoCoMaTech)". HFQ® is a registered trademark of Impression Technologies Limited. Impression Technologies Limited is the sole licensee for the commercialisation of the HFQ® technology from Imperial College London.

Author Contributions: Kailun Zheng and Jianguo Lin conceived and designed the experiments; Kailun Zheng and Junyi Lee performed the development of the mathematical model; Kailun Zheng and Wenchao Xiao performed the experiments; and Baoyu Wang contributed materials and tool design and machining. Kailun Zheng and Junyi Lee contributed to writing this paper. All authors contributed to analyzing the data.

Conflicts of Interest: The authors declare no conflict of interest.

References

1. León, J.; Luis, J.C.; Fuertes, P.J.; Puertas, I.; Luri, R.; Salcedo, D. A proposal of a constitutive description for aluminium alloys in both cold and hot working. *Metals* **2016**, *6*, 244. [CrossRef]
2. Zhang, S.; Chen, T.; Zhou, J.; Xiu, D.; Li, T.; Cheng, K. Mechanical properties of thixoforged in situ Mg_2Si_p/AM60B composite at elevated temperatures. *Metals* **2018**, *8*, 106. [CrossRef]
3. El Fakir, O.; Wang, L.; Balint, D.; Dear, J.P.; Lin, J.; Dean, T.A. Numerical study of the solution heat treatment, forming, and in-die quenching (HFQ) process on AA5754. *Int. J. Mach. Tools Manuf.* **2014**, *87*, 39–48. [CrossRef]
4. Zheng, K.; Lee, J.; Lin, J.; Dean, T.A. A buckling model for flange wrinkling in hot deep drawing aluminium alloys with macro-textured tool surfaces. *Int. J. Mach. Tools Manuf.* **2017**, *114*, 21–34. [CrossRef]
5. Nishibata, T.; Kojima, N. Effect of quenching rate on hardness and microstructure of hot-stamped steel. *J. Alloys Compd.* **2013**, *577*, S549–S554. [CrossRef]
6. Sjöström, J.; Bergström, J. Thermal fatigue in hot-working tools. *Scand. J. Metall.* **2005**, *34*, 221–231. [CrossRef]
7. Maeno, T.; Mori, K.I.; Fujimoto, M. Improvements in productivity and formability by water and die quenching in hot stamping of ultra-high strength steel parts. *CIRP Ann. Manuf. Technol.* **2015**, *64*, 281–284. [CrossRef]
8. Cortina, M.; Arrizubieta, I.J.; Calleja, A.; Ukar, E.; Alberdi, A. Case study to illustrate the potential of conformal cooling channels for hot stamping dies manufactured using hybrid process of laser metal deposition (LMD) and milling. *Metals* **2018**, *8*, 102. [CrossRef]
9. Abdollahpoor, A.; Chen, X.; Pereira, M.P.; Xiao, N.; Rolfe, B.F. Sensitivity of the final properties of tailored hot stamping components to the process and material parameters. *J. Mater. Process. Technol.* **2016**, *228*, 125–136. [CrossRef]
10. Mendiguren, J.; Ortubay, R.; De Argandoña, E.S.; Galdos, L. Experimental characterization of the heat transfer coefficient under different close loop controlled pressures and die temperatures. *Appl. Therm. Eng.* **2016**, *99*, 813–824. [CrossRef]
11. Li, H.; He, L.; Zhang, C.; Cui, H. Research on the effect of boundary pressure on the boundary heat transfer coefficients between hot stamping die and boron steel. *Int. J. Heat Mass Transf.* **2015**, *91*, 401–415. [CrossRef]

12. Chang, Y.; Tang, X.; Zhao, K.; Hu, P.; Wu, Y. Investigation of the factors influencing the interfacial heat transfer coefficient in hot stamping. *J. Mater. Process. Technol.* **2016**, *228*, 25–33. [CrossRef]
13. Liu, X.; Fakir, O.; El Meng, L.; Sun, X.; Li, X.; Wang, L. Effects of lubricant on the IHTC during the hot stamping of AA6082 aluminium alloy: Experimental and modelling studies. *J. Mater. Process. Technol.* **2018**, *255*, 175–183. [CrossRef]
14. Hu, P.; Ying, L.; Li, Y.; Liao, Z. Effect of oxide scale on temperature-dependent interfacial heat transfer in hot stamping process. *J. Mater. Process. Technol.* **2013**, *213*, 1475–1483. [CrossRef]
15. Merklein, M.; Lechler, J.; Stoehr, T. Investigations on the thermal behavior of ultra high strength boron manganese steels within hot stamping. *Int. J. Mater. Form.* **2009**, *2*, 259. [CrossRef]
16. Ying, L.; Gao, T.; Dai, M.; Hu, P. Investigation of interfacial heat transfer mechanism for 7075-T6 aluminum alloy in HFQ hot forming process. *Appl. Therm. Eng.* **2017**, *118*, 266–282. [CrossRef]
17. Xiao, W.; Wang, B.; Zheng, K.; Zhou, J.; Lin, J. A study of interfacial heat transfer and its effect on quenching when hot stamping AA7075. *Arch. Civ. Mech. Eng.* **2018**, *18*, 723–730. [CrossRef]
18. Liu, X.; Ji, K.; Fakir, O.; El Fang, H.; Gharbi, M.M.; Wang, L. Determination of the interfacial heat transfer coefficient for a hot aluminium stamping process. *J. Mater. Process. Technol.* **2017**, *247*, 158–170. [CrossRef]
19. Bai, Q.; Lin, J.; Zhan, L.; Dean, T.A.; Balint, D.S.; Zhang, Z. An efficient closed-form method for determining interfacial heat transfer coefficient in metal forming. *Int. J. Mach. Tools Manuf.* **2012**, *56*, 102–110. [CrossRef]

MDPI

St. Alban-Anlage 66

4052 Basel

Switzerland

Tel. +41 61 683 77 34

Fax +41 61 302 89 18

www.mdpi.com

Metals Editorial Office

E-mail: metals@mdpi.com

www.mdpi.com/journal/metals

www.ingramcontent.com/pod-product-compliance
Lightning Source LLC
Chambersburg PA
CBHW051904210326
41597CB00033B/6013